Pest Management in Store Grains

Pest Management in Store Grains

Editors :

P. NARAYANASAMY
Professor & Head, Dapartment of Entomology
Faculty of Agriculture, Annamalai University
Annamalai Nagar – 608 002, Tamil Nadu, India

S. MOHAN
Professor of Entomology, Department of Cotton
Centre for Plant Breeding and Genetics
Tamil Nadu Agricultural University
Coimbatore – 641 003, Tamil Nadu, India

J.S. AWAKNAVAR
Professor of Entomology, Department of Entomology
University of Agricultural Sciences, Dharwad – 580 005
Karnataka, India

SATISH SERIAL PUBLISHING HOUSE
403, Express Tower, Commercial Complex, Azadpur, Delhi-110033 (India)
Phone : 011-27672469, 27672852, Fax : 91-11-27672046
E-mail :info@satishserial.com, hkjain1975@yahoo.com
Website : www.satishserial.com

Published by :

SATISH SERIAL PUBLISHING HOUSE
403, Express Tower, Commercial Complex, Azadpur, Delhi-110033 (INDIA)
Phone : 011-27672852 Fax : 91-11-27672046
E-mail : info@satishserial.com, hkjain1975@yahoo.com

Print Again : 2023

ISBN : 81-89304-62-3

Composed, Designed & Printed in India

ABOUT THE EDITORS

DR. P. NARAYANASAMY

- Presently serving as Professor and Head, Department of Entomology, Faculty of Agriculture, Annamalai University, Annamalai Nagar, Tamilnadu and has rendered 34 years of service at Annamalai University in the Department of Entomology. He conducted scores of research projects funded by Tamil Nadu State Council for Science and Technology, Government of Tamil Nadu, Ministry of Environment & Forests, Government of India and University Grants Commission, New Delhi.
- At present, two mega research projects on Development and Utilization of Flyash as a Carrier in chemical and herbal insecticide formulations funded by Ministry of Coal and Ministry of Science and Technology, Government of India respectively are underway.
- He has to his credit publication of more than 100 research papers and 8 books and organized two national level seminars on 'Use of Flyash in Agriculture and a Refresher Course on 'Organic Pest Control' during 2008. His fields of specialization are:

❑ Developed a fungal insecticide, Mycoinsecticide 70% WP against rice brown planthopper and green leafhopper

❑ Discovered flyash as an insecticide and a carrier in chemical and herbal insecticide formulations

❑ Documented more than 1000 indigenous pest control practices and revalidated important practices for farmers' use. Also large number of tribal pest control practices collated from various hillocks in Tamil Nadu were also documented and published.

DR. S. MOHAN

- Presently serves as Professor of Agricultural Entomology at TNAU, Coimbatore. He was an awardee of many contributions viz., Tamil Nadu Scientists Award for 2003, Jawaharlal Nehru Award for best Ph.D. thesis on Stored Product Entomology and TNAU's Best Researcher Award for the year 2005.

- He has designed a number of insect traps for storage pests and obtained an Indian patent on a device to remove eggs from stored pulses. He has organized a series of four training programmes on Storage Entomology. He has to his credit publication of 95 research papers, 2 books and 3 book chapters. He is a TNAU approved consultant for entrepreneurs and private warehouse managements.
- His field of specialization is Stored Product Entomology.

DR. J.S. AWAKNAVAR

- Presently serves as Professor of Entomology at College of Agriculture, University of Agricultural Sciences, Dharwad, Karnataka and earlier worked as a Head, Department of Entomology, Dharwad, during 2004-2007. He has to his credit publication of more than 50 research papers earlier he was Principal Investigator of an ICAR Project on Evaluation of toxicity of agrochemicals to earth worms completed in 1998 and has guided good number of candidates for award of M.Sc. (Ag.) and Ph.D. in Entomology.
- His fields of specialization are: Toxicology, Insect Physiology and Storage Entomology.

FOREWORD

Storage of food grains and produce has been a routine practice undertaken by the farm folks at their dwelling places. Certain public sector undertakings namely, Food Corporation of India, Central and State Warehousing Corporations stock large quantities of the food grains. Pests, diseases and moulds pose threat to storage of the grains as their damage would result in significant loss and wastage of the food grains. In the light of this situation, the Department of Entomology, Annamalai University conducted a National Symposium under the title, "Towards pest free grains and seeds in storage" during 26 & 27, March 2007. The meet was a maiden effort in India, had participation of stalwarts of storage entomology nation-wide, and national grain storage organisations viz., Food Corporation of India and Central Warehousing Corporation Ltd.,

I feel happy to learn that proceedings of the first ever symposium on storage entomology are brought out as a book. This contains 31 papers which cover areas like grain storage, pest problems, use of plant products and other eco-friendly pest control strategies.

I find most of the papers interesting and the contents would point to useful manipulation in storage of grains in a better way.

I appreciate Prof. P. Narayanasamy for taking this lead to organize the meet and efforts to bring out this book under title, "Pest Management in Store Grains", and my kudos are due to the editors in getting the papers well trimmed.

I trust that the book will be a useful reference guide for grain storage workers, academia and post graduate students of Entomology.

Sd/-

07th July, 2008 **(T.P. RAJENDRAN)**

PREFACE

Grains in storage either in large or small stock get depredated by varieties of insect pests warranting suitable ameliorative measures including fumigants. Use of such chemical toxicants is on the rise now. Considering this, an attempt has been made to make a congregation on "Towards Pest - Free Grains and Seeds in Storage" at the Department of Entomology, Annamalai University during March 26-27, 2007. This is a maiden meet in India with participation from Tamil Nadu Agricultural University, Coimbatore, University of Agricultural Sciences, Dharwad, Indian Institute of Crop Processing Technology, Thanjavur, Central Food Technological Research Institute, Mysore besides large grain storage organizations like Food Corporation of India (FCI) and Central Warehousing Corporation (CWC).

The symposium became a reality owing to approval by the authorities of Annamalai University for whom our sincere thanks are due.

Organization of the symposium was greeted and blessed by the respected Founder Pro-Chancellor of Annamalai University, Dr. M.A.M. Ramaswamy, Annamalai University. We salute him for the patronage. It is with immense pleasure we think sincerely the respected Vice-Chancellor, Dr. M. Ramanathan for his benevolent support and encouragement in the conduct of the symposium and keen interest in bringing out this book.

Also we feel elated to owe our thanks to our former Vice-Chancellor of Annamalai University, Dr. L.B. Venkatrangan for his constant encouragement and guidance. Our grateful regards are due to Dr. M. Rathinasabapathy, Registrar, Annamalai University for extending all facilities and support in the realization of the meet.

Personally we acknowledge the helpful encouragement and for extending foreword to this book by Dr. T.P. Rajendran, Assistant Director General (Plant Protection) Indian Council of Agricultural Research, New Delhi.

Also we deem it pride to thank Dr. Pritam Chandra, Assistant Director General (PE), ICAR and Dr. R.T. Patil, Director and Dr. S.K. Nanda, Project Co-ordinator, National Centre of Post Harvest Engineering & Technology, Ludhiana for the encouragement. Let me express my appreciation and co-operation rendered by authors of the papers included in this book.

At last, continuous support and encouragement extended by the colleagues at the Department of Entomology, Annamalai University and other organizations is gratefully remembered.

We must express my heartfelt thanks and kudos to M/s Satish Serial Publishing House, New Delhi for bringing out this book in a nice way.

Annamalainagar
09.08.2008

P. NARAYANASAMY
S. MOHAN
J.S. AWAKNAVAR

CONTENTS

Insects Identification and Biology

Insect Pest Management in Storage

Pest Management - Fumigation

Pest Management - Plant Products

Pest Management – Insecticides

Pest Management - Host Plant Resistance

Pest Management - Eco - Friendly Tactics

Non-Insect Pest Management

Quarantine Aspects of Grain Importation

1

Scenario of Pest Management in Grains in Storage

NADARAJAN, L.
Dean, Pandit Jawaharlal Nehru College of Agriculture and Research Institute, Karaikal, Puducherry : Union Territory.

SUMMARY

The paper throws light on the overall feature and status of research on storage pest control in India.

Key words : Fumigants, External Feeder, Transgenic Plants.

1. INTRODUCTION

India is a developing country with a population of 110 crores. Green revolution has increased our productivity *viz.*, 200 million tonnes of food grains, 25 million tonnes of oil seeds and 15 million tonnes of fibre per annum. But, on an average about 7 percent of loss is estimated in storage. Out of this, insects account for 2.5 per cent and rodents damage another 2.5 per cent. Moreover the resistance to common insecticides is more prevalent in major stored grains pests and this is considered as a global phenomenon. Even in early sixties this was witnessed and currently about 31 insect species are found to be resistant to more than 100 pesticides. Among the stored grain insect pests, the red flour beetle, *Tribolium castaneum* (Herbst) is the most common species, to have developed resistance to malathion and lindane in atleast 75 countries of the world. It has also become resistant to phosphine, the widely used grain fumigant. There are also reports to confirm the development of resistance in pea weevil, *Sitophilus oryzae* L. and lesser grain borer, *Rhizopertha dominica* Fab. to malathion/ lindane and phosphine insecticides. Saw toothed beetle, *Oryzaephilus surinamensis* L. developed resistant to phosphine and leather beetle, *Dermestes maculatus* had developed resistance to lindane. A special feature of resistance in the stored grain insect is that they can travel globally through international trade.

2. PEST PROBLEMS IN STORE GRAINS

If we look into the details of insects attacking stored products, these pests may be grouped into internal feeders and external feeders. Among them, the rice weevil, S. *oryzae* is a universally known pest. The other important pests like pulse beetle (*Callosobruchus chinensis* L.) are notorious. The secondary attacking insects which are otherwise referred to as external feeders also cause considerable damage and make the stored products unusable. Among them the red flour beetle, *T. castaneum,* the Indian meal moth, *Plodia interpunctella* and the rice moth, *Corcyra cephalonica* are important. So the protection of stored grains has become very important, since all the attempts made to produce such large quantity of grains and vegetables are destroyed by such pests. So, protection of these grains has become a problem of national importance and can be tackled by two ways: prevention of the attack and control of the stored pests.

Another important major pest, namely rodents cause considerable damage to grains in storage. Among 136 species of rodents recorded in India, only about 16 per cent can be considered economically important. Among them the mole rat, *Bandicota kenodonsis,* the *Ramus raffus,* the brown rat *Rattus norvegicus* Berk., the house mouse, *Mus musculus* L. and the bandicoot rat, *Bandicota indica* Bech. are very important economically, because, of their damage to the stored products.

It is important that the moisture content of the grains should be minimized below 9 to 10 per cent to avoid infestation by majority of the storage pests. So, in a country like India this may be easily achieved by sun-drying and in some countries, they use grain driers. Like humidity reduction, the percentage of oxygen available in the storage structures may also be reduced considerably by various means.

The modern technology of food processing and packaging techniques are taking care of the issues considerably. As a general routine practice, cleaning of godowns, fumigation and cleaning of the storage bags by dipping them in different insecticide solutions are adopted.

3. STORAGE PEST MANAGEMENT STRATEGIES

Research has been conducted on various storage structures of different capabilities to prevent the infestation of rats, insects and microorganisms.

The recent focus for storage pest management is on the use of botanicals against the pests. Botanicals like leaves of neem *(Azadirachta indica* An.dr.Juss.) and Notchi (*Vitex negundo* L.) were used traditionally in Indian villages from time immemorial. Research in this direction has unravelled many potential botanicals on the storage pest management. Few examples are powdered rhizome of sweet flag *(Acorus calamus* L.), essential oil of *Blumea malcolmii,* leaves and stem of *Ocimum sanctum* L., black pepper extract (*Piper nigrum* L.), curry leaf *(Murraya koenigii* L.) (ovicidal), powders of *Ipomoea carnea* L. and *Melia azedarach* L..

In controlling the storage pests, the biotechnological intervention has got good potential. So, a good amount of research is being carried out to utilize the genes responsible for protease inhibition, genes responsible for production of secondary metabolites and use of antisense RNA technology in combination with gene transfer is used for post harvest management. Transgenic plants can thus be used in manipulating enzyme synthesis for the protection of the stored products and also stored vegetables like tomato and ripening fruits.

— xxx —

2

Loss of Grains Due to *Sitophilus oryzae* L. in Different Pop Sorghum Varieties in Comparison with Grain Sorghum

KAVITA JADHAV[1] AND SHEKHARAPPA[2]
[1] Research Scholar [2] Professor
AICSIP, Sorghum Research Scheme, University of Agricultural Sciences
Dharwad-580005, Karnataka

SUMMARY

Three varieties in each of pop sorghum Talakal-6, Mugad local and Shiggaon local and grain sorghum DSV-3, CSH-14 and CSH-16 were evaluated to know the reaction to the rice weevil, *Sitophilus oryzae* L. under laboratory condition. The maximum per cent grain damage of 99.36 was observed with Shiggaon local and minimum of 55.60 damage in DSV-3 at 180 days after storage. Similarly maximum weight loss of 19.30% and maximum population build up of the pest (142.78) was observed in Shiggaon local while DSV-3 recorded least weight loss (5.80%), and minimum population build up (39.82) at 180 days after storage. Out of six genotypes tested grain sorghum variety DSV-3 was found to be relatively resistant to the rice weevil in storage. However among pop sorghum varieties minimum loss and damage was noticed in Mugad local.

Key words: Progeny, Pop Sorghum, Genotypes.

1. INTRODUCTION

Pop sorghum is traditionally grown in central and peninsular India. Majority of them originated in India and popped sorghum grains consumed in all states of India by the poor as well as rich people as snack food and as delicacy. Popped sorghum is considered to be superior to popped corn as it is tender, with less hull, does not clog the space between the teeth. Popped sorghum grains have been found to have as much flavour and is as nutritious as pop corn.

The rice weevil, *Sitophilus oryzae* L. is a serious pest of various food grains under storage. It is largely responsible for damage and frequently harbouring in storage, mills and warehouses. The weevil is of cosmopolitan nature, attributing 100 per cent loss in weight of grains in storage. Post harvest loss was observed from 9 to 20 per cent due to insect pests and microorganisms. The climatic conditions after harvest are favourable to the infestation of various pests by which losses may occur to the extent of 10-15 per cent. *S. oryzae* attacks severely on pop sorghum grains during storage. As the grains need to be stored before popping, identification of resistant pop sorghum genotypes to save from the losses in storage due to *S. oryzae* has assumed greater significance. So in the present investigation some of genotypes of pop sorghum and grain sorghum were screened for their resistance to *S. oryzae*.

2. MATERIALS AND METHODS

Three pop sorghum varieties and three genotypes of grain sorghum were collected and evaluated to assess their reaction to the rice weevil, *S. oryzae* under laboratory condition. Sound, unaffected grains of each variety were dried in hot air oven for 6 hours at 42^{0}C in order to eliminate the infestation by stored grain pests. The moisture content of grains was 10.7 $\pm$ 2 per cent. Grains of each variety weighing 100 g were kept in plastic bottle of 500 ml capacity and replicated four times. Ten pairs of five day old weevils were introduced in each bottle and tops were kept covered with muslin cloth. For every month, each bottle was examined periodically to note the loss of weight of grains, percentage of damaged grains, per cent germination and population build up. Observations were made upto 180 days.

3. RESULTS AND DISCUSSION

The per cent grain damage at 30 DAS ranged from 30.70 to 82.50 in DSV-3 and Shiggaon local, respectively (Table-1). The grain damage increased as the storage period increased. However, at 180 DAS, per cent grain damage was maximum (99.36) in Shiggaon local (Table 1) followed by Talakal-6 (86.70). Grain damage was low in DSV-3 and CSH-16 being 55.60 and 70.90 per cent, respectively. At 180 days after storage, per cent weight loss (Table 2) was more in Shiggaon local and Talakal-6 having 19.30 and 16.50 per cent, respectively. Less per cent of weight loss was noticed in DSV-3 (5.80) and CSH-16 (10.10).

Table 1 : Screening of pop sorghum varieties in comparison with grain sorghum varieties against rice weevil damage

Genotypes	Per cent grain damage						
	30 DAS	60 DAS	90 DAS	120 DAS	150 DAS	180 DAS	Mean
Pop Sorghum:	67.60b	75.70b	80.10b	82.70b	84.50b	86.70b	
Talakal-6	(55.35)	(60.48)	(63.52)	(65.44)	(66.78)	(68.58)	79.55b
Mugad local	50.60c (45.37)	57.80cd (49.47)	61.30d (51.53)	63.40c (52.79)	64.60c (53.50)	71.00c (57.39)	61.45d
Shiggaon local	82.50a (65.30)	88.40a (70.11)	92.70a (74.32)	95.90a (78.30)	97.40a (80.79)	99.36a (85.57)	92.71a
Grain Sorghum: CSH-16	43.90d (41.15)	51.10d (45.64)	56.70d (48.82)	59.70c (50.65)	67.30c (55.28)	70.90c (57.39)	58.26e
CSH-14	56.80c (48. 90)	64.20c (53.28)	72.70c (58.49)	78.50b (62.40)	82.90b (65.58)	86.00b (68.09)	73.51c
DSV-3	30.70e (33.66)	33.10e (35.12)	38.50c (38.343)	44.00d (41.54)	49.30d (44.69)	55.60d (48.21)	41.86f
CD (1%)	3.555	3.875	3.872	3.458	4.388	3.751	0.316
SE m±	0.853	0.929	0.929	0.829	1.053	0.900	0.075

DAS - Days after storage
Means followed by the same letter do not differ significantly by DMRT (P=0.01)
Values in the parentheses are arc sine transformed values.

Highest progeny were (Table 3) produced on Shiggaon local (142.78) followed by Talakal-6 (125.23), while it was least in DSV-3 (39.82) followed by Mugad local (109.54) at 180 days after storage. Out of six genotypes tested grain sorghum variety DSV-3 was found to be relatively resistant to rice weevil in storage. However, among pop sorghum varieties minimum loss and damage was noticed in Mugad local. All these varieties need to be protected in order to avoid loss.

Table 2 : Weight loss of different genotypes of pop sorghum and grain sorghum due to *S. oryzae* infestation

Genotypes	Per cent weight loss						
	30 DAS	60 DAS	90 DAS	120 DAS	150 DAS	180 DAS	Mean
Pop Sorghum Talakal-6	4.60b (12.59)	6.20b (14.40)	8.30b (16.76)	10.50b (18.94)	12.30b (20.50)	16.50b (24.08)	9.73b
Mugad local	2.80d (9.58)	3.80d (11.31)	5.20d (13.24)	7.30d (15.72)	9.40d (17.88)	10.40d (18.91)	6.48d
Shiggaon local	6.60a (14.89)	7.90a (16.29)	9.90a (18.34)	12.20a (20.47)	15.30a (23.02)	19.30a (26.03)	11.86a
Grain Sorghum CSH-16	2.00e (8.13)	3.40e (10.67)	4.9e (12.72)	6.30e (14.51)	7.90e (16.35)	10.10d (18.56)	5.76e
CSH-14	3.20c (10.35)	5.40c (13.45)	7.50c (15.89)	9.40c (17.84)	11.20c (19.46)	14.30c (22.19)	8.50c
DSV-3	0.63f (4.55)	1.70f (7.51)	2.20f (8.54)	3.40f (10.55)	4.50f (12.32)	5.80e (13.92)	3.03f
CD (1%)	0.39	0.16	0.41	0.31	0.51	0.38	0.28
SE m±	0.09	0.66	0.10	0.10	0.21	0.09	0.07

DAS - Days after storage
Means followed by the same letter do not differ significantly by DMRT (P=0.01)
Values in the parentheses are arc sine transformed values.

Table 3 : Population buildup of *S. oryzae* in different genotypes of pop sorghum and grain sorghum

Genotypes	Population build up at						
	30 DAS	60 DAS	90 DAS	120 DAS	150 DAS	180 DAS	Mean
Pop Sorghum Talakal-6	74.46b (8.66)	85.24a (9.27)	89.94a (9.52)	98.10ab (9.93)	113.34ab (10.67)	125.83b (11.25)	97.81a
Mugad local	56.17c (7.53)	77.82a (8.86)	80.50a (9.00)	90.51b (9.54)	102.31c (10.14)	109.54c (10.49)	86.14b
Shiggaon local	79.29a (8.93)	85.61a (9.28)	88.42a (9.44)	101.70a 10.11)	122.26a (11.09)	142.78a (11.98)	103.34a
Grain Sorghum CSH-16	27.30e (5.27)	41.94b (6.52)	55.15b (7.46)	60.65c (7.83)	64.30d (8.06)	112.28c (10.62)	60.27d
CSH-14	38.22d (6.22)	47.66b (6.94)	53.22b (7.33)	92.23b (9.63)	109.54bc (10.50)	114.6c (10.74)	75.91c
DSV-3	14.29f (3.85)	19.75c (4.51)	23.41c (4. 90)	26.74d (5.22)	29.75e (5.50)	39.82d (6.36)	25.62e
CD (1%)	0.20	0.46	0.54	0.41	0.46	0.42	6.36
SE m±	0.05	0.11	0.13	0.10	0.11	0.10	1.53

DAS - Days after storage
Means followed by the same letter do not differ significantly by DMRT (P=0.01)
Values in parentheses are arc sine transformed values.

There is no evidence to make critical discussion on this issue but Borikar and Tayde (1979) reported that hybrids comparatively had less incidence of rice weevil than the local genotypes. This may be attributed to genotypic character having thick pericarp. The damage due to *S. oryzae* affected the viability of genotypes. There was considerable loss in the seed viability (Table 4) at 180 days after storage. DSV-3 recorded maximum germination percentage of 69.00 followed by CSH-16 (59.80).

Table 4 : Per cent germination in different pop and grain sorghum genotypes due to *S. oryzae* infestation

Genotypes	Per cent germination at						
	30 DAS	60 DAS	90 DAS	120 DAS	150 DAS	180 DAS	Mean
Pop Sorghum Talakal-6	60.00e (51.00)	58.80e (50.07)	53.00e (47.70)	50.60e (45.31)	48.40d (44.07)	44.30e (41.73)	52.51e
Mugad local	68.40c (55.80)	66.30c (54.50)	64.30c (53.31)	60.30c (50.92)	56.10c (48.51)	55.20c (47.97)	61.76c
Shiggaon local	45.00f (42.13)	40.10f (39.27)	38.90f (38.58)	35.30f (36.45)	32.30e (34.60)	30.30f (33.41)	37.15f
Grain Sorghum CSH-16	74.00b (59.33)	70.00b (56.79)	67.00b (54.95)	65.60b (54.12)	62.30b (52.11)	59.80b (50.66)	66.45b
CSH-14	64.90d (53.67)	61.00d (51.35)	58.70d (50.05)	55.40d (48.11)	48.40d (44.07)	48.30d (44.02)	56.11d
DSV-3	88.00a (69.73)	80.00a (63.44)	78.80a (62.59)	74.60a (59.76)	70.60a (57.15)	69.00a (56.16)	78.93a
CD (1%)	0.87	0.67	1.35	1.02	0.92	1.98	3.17
SEm±	0.22	0.17	0.35	0.26	0.24	0.51	0.76

DAS - Days after storage
Means followed by the same letter do not differ significantly by DMRT (P=0.01)
Values in the parentheses are arc sine transformed values

Loss of viability could be due to damage caused by *S. oryzae* during storage period. It draws the support of Kurdikeri *et al.* (1993) who reported that initial germination of 97.10 per cent in maize hybrid, Deccan 103 was brought down to 39.50 per cent, after six months of storage due to rice weevil infestation.

References

Borikar, D. S.and D. S Tayde, 1979. Resistance in sorghum to *Sitophilus oryzae* Linn. *Proceedings of the Indian Academy of Science Bulletin.*, 28 : 273-276.

Kurdikeri, M. B., B. Aswathaih, Rajendra Prasad, R.D. Katagal, and S.C. Aswhathanarayana, 1993. Studies on relative seed damage loss in weight and loss in viability caused by *S. oryzae* Linn. and *Rhizopertha dominica* Fab. in maize hybrids. *Seed Res.*, 21: 72-77.

— xxx —

3

An Insight into Traditional Grain Storage in India

NARAYANASAMY, P.
Professor and Head, Department of Entomology, Faculty of Agriculture, Annamalai University, Annamalainagar-608 002, Tamil Nadu.

SUMMARY

Taking cognizance of the pesticidal residues and their effects on the environment and the human being evident as of today, concerted efforts were on documentation, revalidation and application of traditional practices evolved by the rural and tribal folks. These native practices have been analysed and prescribed here in the national scenario so as to usher in relatively pesticide free grain in storage.

Key words: Traditional Grain Storage, Revalidation, Storage Structures

1. INTRODUCTION

The word storage means storing of goods by a particular method. The receptacles used for storage of food grains by the farmers are of different types, shapes, materials and capacities. They are made of the locally available materials like clay, wood, bamboos, straw, jute bags, brick etc., and it has been observed to keep the seeds in good condition for longer periods or till the next harvest. The reason being that this can't protect the grains against pests, moisture etc., in India, out of the total production, about 70% is retained and stored by farmers for consumption, seed and payment of wages. Comparatively higher quantity of grains stored by farmers and traders is not on sound basis and the extent of loss is on the higher side.

Around the world, there is growing interest in finding alternatives to the modern farming methods that have emerged during the 20th century. The deleterious effect of the pesticides, chemical fertilizers and genetic monocultures have become increasingly apparent. One approach is to build upon traditional methods, which got evolved over the first 10, 000 years of agriculture. This had evolved tremendous variety of domesticated crops and livestock, and systems of farming. Fortunately, there is still a vast scope of farming know - how in India. Many traditional farmers in the developing world are still practicing farming methods that are in balance with the surrounding ecosystems stable, sustainable and highly efficient. Such traditional wisdom can perhaps enable the developed work to grow food, slow erosion, control pests and decrease our dependence on chemical inputs.

1.1. Traditional knowledge in Indian Agro - ecosystem can be classified into the following

- Traditional wisdom during shifting cultivation
- Traditional wisdom during Nomadic pastoralism
- Traditional wisdom during continuous cultivation
- Traditional wisdom during mixed subsistence farming

1.2. Following characteristics of the traditional knowledge can be considered important

- Focus our risk reduction
- Year round vegetative cover of soil
- *System diversity :* Farming system based on several cropping system, based on a mixture of crops and crops with varietal and other genetic variabilities
- Trophic complexity approaching natural system. Multiple interactions between (plants, weeds, pathogens and insects).
- High net energy yields, as the energy inputs are relatively low and low levels of inputs and high degree of self sufficiency.

At rural level, grains are stored mostly from 3 months to 2 years depending on the need. Post harvest losses occur both in quantity and quality. It is generally recognized that the losses in stored grains occur due to insects, rodents and fungi, while the insects migrate from the field at the time of harvest to storage site and the fungus is brought in due to wet condition of the grain. Fungi and insects are often interrelated.

1.3. For the best storage performance, the food grains and oil seeds must be

- Thoroughly cleaned, graded and dried to the safe storage moisture level which is generally 10-12% on wet basis (wb) for food grains in a storage structure and 7- 9% (wb) for oil seeds for 6-12 months of storage.
- Keep the cleaned and dried grains in storage structures which has been properly repaired, cleaned and disinfected .
- Structures should bear the load of grains stored and do not permit contact / exchange with outside humid air. The structure should be constructed in the coolest part of the house / farm. For long term storage the moisture levels are to be kept low.
- The receptacles used for storage of food grains by the farmers are of different types, shapes, materials and capacities. These are made of the locally available materials like clay, wood, bamboos, straw, jute bags, brick etc.

1.4. Existing and proposed structures should be assessed for their suitability for the following functions

- Security of stocks
- Protecting the commodities against damage from rain, flood, damp, solar heating and pests including insects, rodents and birds.
- Accessibility of stocks for accounting, assessment of quality, pest control and remedial action to reduce other forms of deterioration.

While going into the indigenous practices, it is interesting to bring them together as adopted in various parts of our country. Following are various kinds of traditional practices followed in different states of India.

2. INDIGENOUS GRAIN STORAGE PRACTICES

2.1. Tamilnadu (Baskaran and Narayanasamy, 1995; Narayanasamy, 2002; Karthikeyan, *et al.*, 2006)

- Mix dried leaves of neem with grains and pulses
- Place leaves upon jute bags with stored grains / pulses and also mix with them
- Mix red chillies with pulses
- A mixture of cow dung + neem oil is used for plastering mud walls and floor in storerooms
- Bamboo bins are painted with neem cake paste
- Mud pots can be painted with neem cake paste
- Dip the jute bags used for storage in neem seed solution
- Sweet flag rhizome powder can be taken in cloth bags for storage
- Use paddy husk as a dunnage material during storage of paddy grains
- Mix wood ash with paddy seeds
- Mix dried pungam leaves with grains
- Spread neem and pungam leaves at the bottom of mud bin just below mouth level before filling the bin with grains
- Nanchilaiperattai (*Tylophora indica* (Burm.f.) Merr.) rhizome powder is mixed with grains that can be stored upto 30 years
- Tobacco and water is used to control ants
- Mix turmeric powder with the grains
- Custard seed powder is mixed with pulses
- Bins are made of *Vitex* fibre plastered with cow dung and used in grain storage
- Pulse seeds are mixed with ash to absorb excess moisture during storage

2.1.1. Storage of rhizome and tuber

- Pits are formed in a thatched shed with walls coated with cow dung paste (bottom layer sand / sawdust, next layer is of ginger rhizomes and then alternated with sand / saw dust). At the top sufficient gap is provided for aeration and inspected once in 20 days.

- Pits are dug under shade where there is no seeping of water into them.
- Loose heaping over a layer of sand / paddy husk and covered with dry leaves in a thatched shed
- *Lantana* plant leaves ash is used for storing seed potato tubers
- Most of the farmers use *Vitex* leaves to mix with stored products in mud bin or mud sal
- Dry neem leaves are mixed with products in storage
- Dry chillies are mixed with commodities in wooden pathayam for long life
- Dry tulsi leaves are mixed with the cereals to store for long period
- Dry chillies + dry tulsi leaves are mixed together with stored products for storage for long period with quality
- Use ash for mostly vegetable seeds for storing in bins for long period
- Mix red soil with the cereals and pulses for long keeping quality
- Parboiled rice is mixed with rice bran for long time
- Dry turmeric pieces are mixed with pulses and other oil seeds in bins for storing for long period
- Dry sand is mostly mixed with vegetable seeds for storing for long period
- Sundrying of the cereals and pulses is an important practice for long storage
- Some of the farmers are using paddy stalk for covering the stored products in square shape and keep them airtight
- Using mudsal for storing seeds of cereals and covering with surrounding mud and cowdung kept inside with products
- Use of pearl millet or maize stalk as cover for the products and keep them air tight
- Reduce the moisture level of the products kept in the bins for long time keeping quality
- Dry *Adathoda* leaves are mixed with products and kept in bins for long period with quality

- Tanjavur traps are installed at specifically activated places of rats with bait (coconut) catch to the rats
- Gunny bags are used to store for long period
- Mud pots are used to store the products with quality for long time

2.1.2. Rodent control

- Spread a fine layer of ragi over surface of grains stored in bins
- Papaya pieces are kept at the site of movement of rats

2.2. Orissa (Srivastava & Pandey, 2006)

- Mix neem leaves with rice, wheat & other grains
- Dry leaf powder of *Lippia semhata, Aegle marmelos, Ocimum americanum* at 1:100 parts
- Dipping gunny bags in oil of *Acorus* 0.1%
- Paste of young leaves of *Toona hexandra*

2.3. Northern India (Tesfage & Gautam, 2003)

- Pulses are mixed with ash + dung
- Pulses are mixed with ash + chilli
- Pulses are mixed with chilli + salt
- Pulses and cereals are mixed with sand
- Leaf powder of *Croton macrostachynos, Phyllodendron deodecandra,* cuttings of *Aloe* spp, Persian lilac, *Ruta chalepensio*, Mexican marigold, *Datura, Germarium, Lantana camera* are used to mix with the seeds
- Mix seeds with ash + garlic

2.4. Andhra Pradesh

- Sun - drying of pulse seeds is done and packed in gunny bags
- Pulses are packed tightly in double gunny bags or hand knit gunny bags to prevent movement of insects
- Smaller cereal grains are mixed with pulses to fill intergrandular space

- Pulse seeds are placed in a layer of dry sand / cow dung and clay mix at the top (20cm thickness)
- Leaves of neem and pungam are mixed with seeds in bins / bags.
- Pulses are coated with a thin layer of edible oils of neem, pungam etc.,
- Dried pulses are warmed by mixing with wood ash / sand and stored in earthern pots

2.5. Madhya Pradesh (Vinayagam *et al.*, 2006)

- People use leaves of plants like neem, *Vitex, Adathoda, Clerodendron infortunatum* & sweet flag.

2.6. Assam (Deka *et al*, 2006)

- A pot of water is kept over the stored grain
- Keeping neem leaves over the stored rice, mustard oil with stored grains
- Sand is given as a cover (2 - 5 cm depth) over the potato tubers in storage.
- Similar to these practices, the tribal folks have a rich knowledge in this regard.

3. TRIBAL STORAGE PRACTICES IN TAMILNADU (Kathirvelu and Narayanasamy, 2006)

- Mixing pulse seeds with ash
- Mixing ashes, lime and crushed chillies with store grains
- Mixing green gram with sand
- Coating red gram with red earth
- Mixing neem leaves with cereal grains
- Use of nochi leaves for pulse storage
- Sundrying of cereals and pulses
- Mixing table salt with grains
- Use of *Clerodendron* in pulse storage
- Mixing *Acorus calamus* with seeds

- Use of ash and ragi husk for safe storage
- Pulses with Naithulasi *(Ocimum canum* Sims) and chilli
- Mixing pulses with coconut oil
- Pulses are mixed with sand
- Splitting pulses before storage
- Drying pulse grains before storage
- Red earth coating of pulses is a common traditional practice
- Mixing green gram with ash

4. TRADITIONAL STORAGE STRUCTURES

As evidenced by the number of technique of grain storage with no or less pest information tribal folks have also evolved varieties of storage structures small or large depending upon their produce or tenure of storage. Following are some of the important storage devices designed and used by the people indigenously.

4.1. Classification of Storage Structures

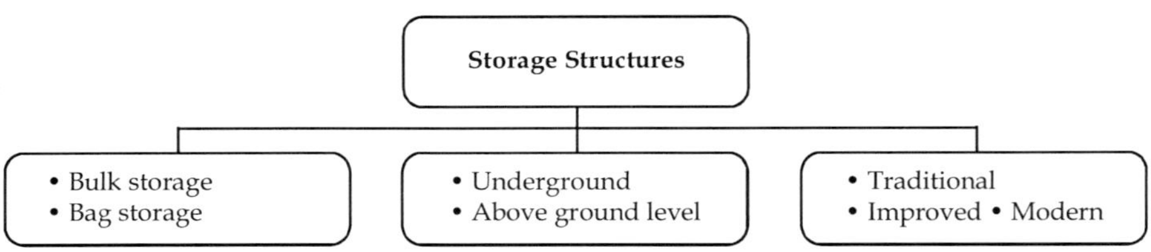

4.2. Normally the traditional storage structures introduced in South India can be divided into two major categories

- Those made of either unburnt clay, stone slabs, bricks etc., i.e. which can be made, reasonably air tight. Such receptacles are used for storage of food grains with low moisture content (i.e.) wheat, peas etc., these receptacles have two major defects of not being rat and moisture proof.
- Those made of bamboo, wood, straw and other dried plant materials locally available with the farmers. It is difficult to make them air tight. Such receptacles are used for storing food grains like paddy, maize, etc.,The perforations in the receptacles aid in allowing the air to move through the grain; thus allowing the

grains to dry to some extent. Such receptacles are generally kept in open to facilitate free circulation of air and ultimately quick drying of grain. This can be placed on raised platform to avoid any damage to the grain due to the sub - soil water. These structures are not proof against insects, rodents, fire hazards, domestic animals, theft, rains, sub - soil water etc. Moreover, the life of these structures is short and are required to be constructed every year.

4.3. Following are certain selected structures adopted at villages

- Mud pot/Sal
- Mud bin
- Bamboo bin
- Wooden pathayam
- Kalanjiyam
- Puri
- Gunny bags
- Underground storage
- Wooden boxes
- Drums/tin
- Containers

4.3.1. Bamboo bin

These structures are commonly used for the storage of paddy, wheat and sorghum in different parts of the country. They are made up of split bamboo woven in the form of a cylinder with wide base and narrow mouth. The bamboo structures are called Thombay ' in Tamil Nadu. They are mainly indoor structures used for storage for 4 - 5 years, and the weight loss due to insect infestation is 5% in paddy and 15% in sorghum. They are kept over the wooden benches (or) fixed with the metal cones at the base.

4.3.2. Mud bin (Kudhir/seru)

These mud structures are made up of good quality clay and mixture of paddy wheat straw and cowdung in the ratio of 3 : 3 : 1. The clay earthen structures are made, sun dried and then burnt in fire. They are common in Tamil Nadu. These are indoor structures used for the storage of paddy, wheat, sorghum, oilseeds and pulses.

The storage capacity varies from 5 to 10 kg. and its life varies from 8-10 years. The extent of weight loss due to insect infestation is 5 - 100 in paddy, 10 - 15% in wheat and 25% in bengalgram. Such structures develop cracks during rainy season. They are also not suitable for fumigation. Seru is used for storing more than three varieties of grains in the same structure. They may be painted with coal tar outside and the base should be constructed with concrete and an outlet be fixed at the bottom

4.4.3. Wooden pathayam

The wooden structures for storage of grains, mostly paddy are known as 'pathayam' in Tamil Nadu. They are made up of locally available wood which is painted black. The capacity of the bin is 10 quintals. At the top, 30cm X 20cm inlet is provided to fill the grains and at the bottom 30cm X 15cm oulet to unload the bin. The bin is 150cm long, 90cm broad and 187.5cm high. It has 4 legs with a height of 45cm. This type of structure is used for storing seeds and food grains especially in the paddy belts of Thanjavur, Cuddalore and Villupuram districts of Tamil Nadu. They are very much durable indoor structures lasting for 15 - 20 years. The loss in weight due to insect pest infestation is only 2-2.5% in paddy. However, in the existing conditions, the bin is neither airtight nor moisture proof. It is also not free from the damage of insects and rodents.

4.3.4. Kalanjiyam

The farmers having large land holding where average storage capacity varies from 50 - 200 quintals, use these indoor structures land are part of the house. They are known as kalanjiyam in Tamil Nadu. They are rectangular, constructed with brick in cement or lime mortar having a wall thickness of 40 - 50cm. They are either terraced or with G. I. sheets. They have compartments for different commodities. An opening of about 50 X 50 cm is provided at the top of the wall for filling the grains and an outlet of about 15cm X 15cm below for taking out the grain. These structures are more or less permanent natured and the expected life is 25 - 30 years. The major commodities stored are paddy, sorghum and wheat, and the weight loss due to insect infestation ranges from 2 - 5%, 15 - 20% and 5 - 6% respectively.

However, the disadvantages are that the initial cost is more and it absorbs moisture from the atmosphere very easily and large surface area of grains is prone for the activity and multiplication of insects.

Possibilities for rat damage are also more as they can move through the roof easily. For better efficiency meshes can be fit, wooden doors provided with metal sheet and cement martar applied on the outer surface.

4.3.5. Puri

There are certain grain storage structures made up of plant materials like paddy straw known as "Puri". This is an outdoor structure where paddy straw would be in the form of rope to varying diameter and capacity of 30-100 quintals. The speciality of this structure is that the construction is followed by loading and it looks as though coiled around the grain. The floor and roof may also covered with straw. The structures may be damaged by insects or rats and the extent of loss will be 15 - 20% in paddy.

This is good for temporary storage immediately after harvest. The cost is very low. The main disadvantage is that no intermittent withdrawals are possible. For better use, a concrete base is provided and an RCC concrete ring wall be constructed around the structure from the floor.

4.3.6. Gunny bags

The gunny bag made of Hessian cloth is the most prevalent storage device all over the country. In Tamil Nadu it forms 77.5% of the storage facility in use. The use of gunny bags facilitates easy handling and accounting of the stock. The cost is less in initial stage and it can be used for multi - various purposes. However, it is not rat, insect and moisture proof. Life of gunny bags is 2 years. The extent of loss due to insect infestation is 2 - 5% in paddy and 8 - 10% in wheat during storage period of one year. For more life, damaged bags need to be changed immediately, avoid storing bags over floor instead of we can go for wooden crates, bags should be stored 1 or 2 ft. away from the walls.

4.3 7. Mud pot / Sal

Earthen receptacles are generally used by the farmers in various sizes and shapes for keeping small quantities of grain. These vessels may be small or big provided with earthen lid to close the vessel when filled with grain. It is better to keep it on wooden bench and it can be painted coal tar.

4.3.8. Underground storage

Farmers construct underground storage structures when they build the house. The mouth size of the under ground is 3 X 3 feet. Black stone commonly called as 'Solakuli' in Coimbatore and Periyar Districts is used as a flooring material. After harvest, grains are stored here. They fill the grains in whenever they need get down inside with the help of a ladder and set them out. We can store 20 - 50 bags here.

4.3.9. Drums, tin bins and tin containers

Oil drums, tin bins and tin containers are also used for storing grains in small scale. The sizes vary according to the need to grain storage. This method is followed in UttarPradesh. Among the above mentioned storage structures, grains stored in gunny bags are highly infected by *Aspergillus flavus* and also it is having highest level of aflatoxin B1.

4.3.10. Traditional storage structures in India (IGSI, 1980)

From the documentation of traditional knowledge, it is known that there are varieties of storage structures which vary from state to state. Following are notable instances.

4.3.11. Andhra pradesh

a) *Puri :* Paddy straw storage structure (capacity 3.0 - 20.0 M.T.)

b) *Kotlu :* Masonry storage structure (capacity 7.5 - 30.0 M.T.)

c) *Gade/Guml/Borem :* Bamboo / Reed: storage structure of the capacity 2.0 - 50.0 M.T. for out door and 0.75-1.0 M.T. for indoor.

d) *Pathara :* Underground storage structure (capacity 1.0-2.0 M.T.)

f) *Basta / Sanchi:* Hessian storage structure (capacity about 75 Kg.)

4.3.12. Tamil nadu

a) *Seru :* Paddy straw storage structure (capacity 5.0 - 15.0 M.T.)

b) *Kottal :* Paddy straw storage structure (capacity 50 - 100 kg.)

c) *Kalanjiam :* Masonry storage structure (capacity 3.0 - 25.0 M.T.)

d) *Pathayam :* Wooden straw storage structure (capacity 0.5 - 2.0 M.T.)

e) *Kudir:* Bamboo / Reed: storage structure (capacity 2.0 - 4.0 M.T.)

f) *Thombe :* Bamboo / Reed: storage structure (capacity 400 - 1000 kg.)

g) *Jadi:* Earthern structure (capacity 50 - 100 kg.)

h) *Koni:* Hessian structure (capacity 75- 100 kg.)

4.3.13. Karnataka

a) *Puri* : Paddy straw storage structure (capacity 3.0 - 20.0 M.T.)

b) *Kanaja* : Masonry storage structure (capacity 1.0-20.0 M.T.)

c) *Panatha* : Reinforced Bamboo structure (capacity 4.0 - 40.0 M.T.)

d) *Ambarkani*: Partly above ground and partly underground structure (capacity 1.0-20.0M.T.)

e) *Thombe/Ponaka/Galag/Gummf:* Bamboo / Reed structure (capacity 2.0 - 4.0 M.T.)

f) *Gade (out door):* Bamboo / Reed structure (capacity 2.0 - 20.0 M.T.)

g) *Vadal/Adakalu/Madeke/Kanaja* : Earthen structure (capacity 5 to 200 kg.)

h) *Hagevu* : Underground storage structure (capacity 1.0 – 8.0 M.T.)

i) *Goni Cheela :* Hessian structure (capacity 75- 100 kg.)

4.3.14 Kerala

a) *Pathayam* : Wooden structure (capacity 0.5 - 2.5 MT.)

b) *Ara* : Wooden structure (capacity 2.5 - 20.0 M.T.)

c) *Vallam* : Bamboo structure (capacity.150 - 500 kg of paddy.)

d) *Mankalam* : Earthern structure (capacity 45-75 kg of paddy and 50-100 kg other cereals.

e) *Chaku* : Messian structure (capacity 75-100 kg.)

4.3.15.Orissa

a) *Dhana* : Straw storage structure (capacity 3.0 - 20.0 M.T.)

b) *Ollya*: Straw storage structure (capacity 35 - 75 kg.)

c) *Amar*: Masonary storage structure (capacity 2.0 -10.0 M.T.)

e) *Dholl*: Bamboo/ Reed: structure (capacity 150 -600 M.T,)

f) *Ghama*: Earthern structure (capacity 50 to100 kg.)

g) *Khar*: Underground storage structure (capacity 1.5 - 20.0 M.T.)

h) *Basta* : Hessian structure (capacity 75 kg.)

4.3.16 *Haryana*

a) *Kaccha kothi/Mud bin:* Mud structure (capacity 250 kg)

b) *Kuthle*: Mud structure of smaller capacity.

4.3.17 *Punjab*

a) *Mud bin/Bharola :* Indoor mud structure (capacity 50 kg to 5 tonnes which outdoor Bharola are of the capacity upto 3.5 tonnes.)

b) *Peru*: Indoor structures made of split bamboo (capacity 100 kg – 1 ton)

c) *Kaccha kothl:* Mud structure (capacity 500 kg – 2 ton)

d) *Theka*: Hessian cloth structure (capacity 500kg-1ton)

4.3.18 *Himachal pradesh*

a) *Peru* : Indoor bamboo structure (capacity 300-700 kg.)

b) *Petti*: Wooden structures in the shape of boxes of capacity 300-700 kg.

4.3.19 *Jammu & Kashmir*

a) *Petti*: Wooden / iron structure of capacity 300 -700 kg.

b) *Kothar*: Outdoor wooden structures in the shape of boxes for higher capacity

4.3.20 *Rajasthan*

a) *Obri/Kuthal/Kuthla :* Mud structure outdoor of capacity 100 -500 kg.

b) *Bukharf*: Outdoor bamboo structure of capacity 1-1.5 ton.

c) *Gare ki Kothl:* Mud structure of smaller capacity

d) *Pacckl Bukharl*: Masonry storage structure of capacity 1 -1.5 ton.

e) *Khanl*: Underground structures of capacity upto 60 ton.

f) *Thakka*: Indoor hessian cloth structure of capacity 800kg-1ton.

4.3.21 Gujarat

a) *Obri/Kuthal/Kuthla :* Mud structure outdoor of capacity 100 -500kg.

b) *Bukharl*: Outdoor bamboo structure of capacity 1-1.5 ton.

c) *Gare ki Kothl:* Mud structure of smaller capacity

d) *Pacckl Bukharl:* Masonry storage structure of 1 -1.5 ton.

e) *Khanl:* Underground structures of capacity upto 60 ton.

f) *Thakka*: Indoor hessian cloth structure of capacity 800kg-1ton

4.3.22 Madhya pradesh

a) *Bindi* : Indoor, mud and stone slab structure of the capacity 100 kg - 3.5 ton.

b) *Mud bin:* Indoor mud structure of the capacity100kg - 2 ton.

c) *Kaccha kothi:* Kuthla: Mud structure of smaller capacity (upto 100 kg).

d) *Banskl Bind:* Indoor bamboo structure used in tribal areas of capacity 1-1.5 ton.

e) *Bag Storage:* Bags of food grains surrounded under the bhusa in a kaccha room, capacity varies from 10 - 20 ton.

f) *Room Storage:* Masonry structure for both bulk and bag storage of capacity upto 20 ton.

g) *Banda*: Outdoor underground structure of capacity 5 to 30 tonnes used for bulk as well as bag storage.

4.3.23 Maharashtra & Goa

a) *Dholl/kanagl :* Bamboo/reed storage structure of the capacity 50 to 100 kg.

b) *Kothl/kangl/Ranjan :* Mud structure of the capacity 100 kg - 1 ton.

c) *Pev*: Outdoor underground structure of capacity upto 10 tonnes.

4.3.24 Uttar pradesh

a) *Kaccha kothi:* Mud structure of the capacity 250 kg.

b) *Earthern pot:* Indoor made from burnt clay in shape of matka, capacity varies from 200 - 500 kg.

c) *Theka* : Indoor, hession cloth structure supported on bamboo post capacity 0.8 -1 ton.

d) *Pett* : Indoor wooden structure of smaller capacity

e) *Bhukharl* : Outdoor, bamboo or Sarkanda structure of capacity 1 -1.5 ton.

f) *Khattl* : Outdoor, underground structure of capacity 6 ton.

4.3.25 Bihar

a) *Kaccha kothi:* Mud structure of the capacity 250 kg.

b) *Bohurl/kunda :* Mud structure made of burnt clay of capacity 250 kg.

c) *Mora* : Outdoor, paddy straw rope, structures of capacity 2 -5 ton.

d) *Bukharl*: Outdoor, cylindrical bamboo/sarkanda structure of capacity 1-1.5ton.

4.3.26 Assam

a) *Bhoral* : Outdoor, bamboo and timber storage structures of capacity 2 – 15 ton

b) *Dull :* Indoor, bamboo storage structure of capacity 100 kg -1ton.

c) *Topa/Tom :* Bamboo storage structure of capacity 10 -30 kg.

d) *Mer:* Bamboo mat structure (indoor) of capacity 500 kg -2 ton.

4.3.27 Arunachal pradesh

a) *Bhoral :* Outdoor, bamboo and timber storage structures of capacity 2 – 15 ton.

b) *Gam:* Wooden structure of capacity 400 kg -1 ton.

c) *Bran :* Cylindrical bamboo storage structure of capacity 200 - 500 kg.

d) *Turn :* Bamboo mat structure of capacity 200 kg -1 ton.

4.3.28 Meghalaya

a) *Bakash :* Indoor, wood structures of capacity 40 - 400 kg.

4.3.29 Nagaland

a) *Bharal*: Outdoor bamboo structure of capacity 2-15 ton.

4.3.30 Manipur

a) *Bakash* : Indoor, wood structures of capacity 40 - 400 kg.

b) *Kal/Bhoai* : Outdoor, bamboo & timber structure of capacity 2–15 ton.

4.3.31 Mizoram

a) *Bakash* : Indoor, wood structures of capacity 40 - 400 kg

4.3.32 Sikkim

a) *Dekuti* : Indoor, wooden structures of capacity 100 kg – 1 ton.

b) *Dola*: Indoor split bamboo structure of capacity 50 - 500 kg.

4.3.33 West Bengal

a) *Moral* : Paddy straw structure of capacity 400 kg – 4 tonnes for outdoor and 100 - 400 kg for indoor.

b) *Gota* : Outdoor, split bamboo structure of capacity 800 kg -8 tonnes for paddy.

c) *Dool* : Indoor bamboo structure of capacity 50 - 400 kg.

d) *Kaccha kothi*: Indoor mud structure of capacity 20 kg – 2 ton.

e) *Earthen Jala*: Indoor clay structure of capacity 5 -100 kg

f) *Kucha* : Indoor bamboo structure of capacity 100 -500 kg.

g) *Pool* : Indoor bamboo structure of capacity 200 -800 kg.

h) *Jabbar*: Indoor, bamboo structure of capacity 600 kg -1.2 ton.

5. GRAIN STORAGE BY TRIBALS

From a series of research, it was learnt that the tribals living in inaccessible regions in the hills are rich sources of indigenous knowledge covering storage of grains (Kathirvelu and Narayanasamy, 2004 : Narayanasamy, 2006).

5.1. Following are certain illustrations from Meghalaya (Jeeva *et al.*, 2006)

- Storage structure is ventilated outdoor structures constructed with locally available materials such as bamboo, linda and thatch grass. This is suitable for high humid conditions.
- Rice is stored in store house with raised wooden or bamboo floor. Rice is stored in specialized structures called 'Thiar' in Khasi. The thiar is a small container closely woven with bamboo split into strips. This is thickly padded inside using rice straw.
- The wooden bin made by the Kharsi tribes is called 'Duli'. It is a double layered bamboo basket, plastered on both the sides with cow dung and mud. This is found best for storing wheat grains.
- The tribes store maize as cobs in unhusked condition. The cobs are fastened on frame made out of bamboo poles and kept at a certain height to prevent from animals. A covering thatch grass is placed on the top to protect the cobs from rain. Sometimes a conical bamboo box is fixed upside down in the poles to keep the rats away from the cobs.

5.2. Himachal Pradesh (Kanwar and Sharma, 2006)

Kothij : A rectangular mud structure for storing wheat, maize, paddy, covered at the top and has a window of 60 X 60cm square.

Kuthla : A small cylindrical / rectangular structure for storing flour & jaggery.

Kulhi & Kulha: Small mud structure for pulses and seeds

Lakolu : It is a fixed storage structure in the centre of a wall of a room, constructed with mud bricks.

Peru : It is a bamboo structure, cylindrical shaped. A special variety of bamboo, 'magar' is used.

Chabri : It is a basket with a lid made of 2.5 cm wide bamboo strips in mat form & circular in shape.

Peti : It is a wooden box for grain storage in hilly areas.

Cott : It is a special storage structure for eatables in farm house.

Sting dibba : It is a small box for storing honey, butter and ghee, cylindrical in shape.

Tunn : It is a wooden structure fixed in the wall and partitioned into two storing two different grains.

Kuthar: It is an outdoor structure made of Deodar *(Cedrus deodara)* wood.

Tandup / Khalari / Khul / Khal. It is a bag made by stitching skin of sheep or goat, used to carry grains while hunting.

5.3. Jammu

A safe storage structure for dry fodder / straw has been in use. It is named as 'Fuhadaha' derived from Doga language means a hut like structure constructed with thatched roof. The stored fodder remains safe during rainy season for many years.

6. CONCLUSION

From the above it can be elucidated that the rich indigenous knowledge of the rural people has incorporated a number of locally available plant species and other substances in keeping their produces in their houses free of pests and in the fabrication of array of storage structures. By adoption of such plant species in warding of the pests, the grains are not impregnated with chemical toxicants and are health proof. The storage structures developed by them are economic and keep the yard free from the pests.

Many such precious knowledge remain unexplored in the country requiring proper documentation and revalidation in a coordinated way. Such feature of learning from the experience will pave way for better grain storage for the future of our country.

"Every seed saved in storage is a better proposition than producing seeds at the fields".

References

Baskaran, V. and P. Narayanasamy 1995. Traditional Pest Control CaterpillarPublications, Mariyappanagar, Annamalainagar - 608 002, Tamilnadu.

Deka, M. K., M. Bhuyan and L.K. Hazarika. 2006. Traditional pest management practices of Assam. *Ind. J. Traditional Knowledge*, 5 (1), pp.75 -78.

IGSI. 1980. Storage structures. Indian Grain Storage Institute, Government of India. Hapur. P44.

Jeeva,S,R.D.N., R.C.Laloo and B.P. Mishra. 2006. Traditional agricultural practices in Mehgalaya, North East India. *Ind. J. Traditional Knowledge,* 5 (1) : 7 – 18.

Kanwar, P. and N. Sharma. 2006. Traditional storage structures prevalent in Himachal homes. *Ind. J. Traditional Knowledge,* 5 (1): 99 - 103.

Karthikeyan, C, D. Veeraragavathatham, D. Karpagam and S. Ayisha Firdouse 2006. Indigenous storage practices in pulses. *Ind. J. Traditional Knowledge,* 5 (1): 104-107.

Kathirvelu, C. and P. Narayanasamy. 2004. Indigenous Pest Control Knowledge of Tribals in Tamilnadu. Pp.- 58 - 67. In: Ignacimuthu, S and S. Jayaraj (Eds.) Sustainable Insect Pest Management. Narosa Publishing House, New Delhi.P298.

Narayanasamy, P. 2002. Traditional pest control. A Retrospection. *Ind. J. Traditional Knowledge,* 1(1): 40-50.

Narayanasamy, P and C. Kathirvelu. 2006. Waning Tribal Wisdom on Pest Control. Caterpillar publications, Mariyappanagar, Annamalainagar - 608 002, Tamilnadu.

Narayanasamy, P. 2006. Traditional Knowledge of tribals in crop protection. *Ind. J. Traditional Knowledge,* 5(1):64-70.

Srivastava, S. K. and H. Pandey. 2006. Traditional knowledge for agro -ecosystem management. *Ind. J. Traditional Knowledge,* 5 (1): 122- 131.

Tesfage Adane and R.D. Gautam. 2003. Traditional pest management practices and lesser expoloited natural products in Ehiopia and India; Appraisal and revalidation *Ind. J. Traditional Knowledge,* 2 {2): 189-201.

Vinayagam.S., B. U. Dupane and O. P. Joshi. 2006. Traditional Technology in soybean cultivation in Madhya Pradesh. *Ind. J. Traditional Knowledge,* 5 (1): 25 -33.

— xxx —

4

Eco-friendly Post-harvest Technologies for Management of Stored Grain Insects at Farm, Home and Warehouse - A Step Towards Clean Grain (Insect Free) Revolution

MOHAN, S.
Professor of Agrl. Entomology, Department of Cotton,
Centre of Plant Breeding and Genetics, TNAU, Coimbatore – 3

SUMMARY

The paper describes various devices designed and released by Tamilnadu Agricultural University, Coimbatore for use in the control of pests in grains in storage. Advantages and state-of-art of the traps as encouraged by certain organizations in the country and abroad are highlighted.

Key words: Post-harvest, Eco-friendly, Grain Revolution

1. INTRODUCTION

Among biotic and abiotic factors which affects grains in storage, insects play a major role in the deterioration of food grains causing both quantitative and qualitative losses. Often the presence of the insects in store houses is felt only when they are hovering and fly around, by which time enormous loss and population build up of insects might have occurred.

It is well proved that no granaries can be filled with grains without insects as the harvested produce contain egg (or) larvae (or) pupae in them because of field carryover infestation which cannot be avoided in developing countries like India. So what is required is simple technologies for timely detection of insects in the stored produce and thereby plan timely control measures.

We have developed many devices for stored grain insects some of them are popularly used across the country in households / farms / godowns.

2. LIST OF GRAIN PROTECTION DEVICES

- TNAU insect probe trap
- TNAU Pitfall trap
- TNAU Two-in-one trap for pulse beetle
- Indicator Device
- Automatic insect removal bin
- UV - Light trap for warehouse
- Stored grain insect pest management kit

All these devices can be used for both monitoring and mass trapping of stored grain insects. It is important to note that even a single live insect presence in the food grain can not be tolerated as it may build up and cause enormous loss in storage due to its high reproductive rate.

3. WHAT IS HAPPENING NOW?

Farmers immediately after harvest dry their produce below 10% MC (moisture content) and store them in bags and bins .The harvested produces contain insect life stages ie. egg, larvae (or) pupae (which are inside the grain) which are carried over from field. It is well proved by researchers that this field carry over damage is <1% but if it is not controlled, total loss can be expected within 2 - 3 months. It is well known that from these field carry over infestations adults generally emerge 15 - 20 days after storage and start mating, laying eggs and thereby multiply into large numbers. These insects have high reproductive rate.

The traditional practice followed by the farmers is "SUN DRYING"

They do it generally after a month of storage of food grains by which time the eggs of the insects hatch and enter the grains and get concealed in many cases. Sun drying can kill only eggs and not the larvae (or) pupae which are concealed inside the grain.

Hence, there is a need to timely detect the emergence of field carry over infestation during storage and to take up timely operations like sun drying which will help to kill the eggs and thereby prevent multiplication.

Here comes the use of our devices in bin (or) bag storage, both for control and detection as 'Zero Tolerance' limit is prescribed for many commodities with reference to insects by many countries for export under WTO Treaty.

4. FOOD WITH HUMAN DIGNITY

Food with human dignity should be the basic approach in the years to come. The poor should not be subjected to a patronage approach and referred to as "beneficiaries" but should be treated as partners in achieving the aim of ensuring that every child, woman and man in the country have an opportunity for a productive and healthy life. The right to adequate healthy food and clean drinking water should be regarded as a basic right. Based on this the Tenth Five Year Plan of our country has shifted the emphasis from Food security at he household level to nutritional security at level of each individual. By shifting the attention to individuals, the strategies adopted will help to foster a life cycle approach in nutrition intervention.

It has been clearly established that stored grain insects cause considerable quantitative and qualitative loss to the grains and indiscriminate use of fumigants results in pesticide contamination in the grains and development of insect resistance.

According to the Prevention of Food Adulteration Act, GOI, the amount of weevilled grain (insect damage) (by count) shall not be more than 10 per cent (or) Uric acid content arising as a result of the insect damage shall not exceed 10 milligram per 100 gram whichever is less.

All the above points are very important to satisfy the Tenth Five year plan thrust of 'Nutritional Security' at individual level.

Our contributions will thus help in the betterment of our society. "Every Indian should eat grains with good quality like western people where "Zero tolerance level" is prescribed for the insects attacking grain.

Encourage entrepreneurship to achieve economic growth is the current slogan for the prosperity of our country: Our technologies have created an entrepreneur who is well pushing by way of marketing the technologies across the country. Details are discussed in the impact statement presented in the later part of achievements.

5. KINDER THE SCIENTIFIC TEMPER OF SCHOOL CHILDREN

Recently these technologies have made significant impact in the minds of the school students of Tamil Nadu which has resulted in the development of new innovative ideas among them leading to development of new models by the students. Some of these models have won the National Awards by the students. This, we feel as a significant contribution as our technologies have sown the seeds of innovative ideas among the school children. This will pave way for development of many more newer technologies when they grow which will help to make a *Second Green Revolution.*

Precisely my work try to fulfill the following objectives which are proposed in the 10th plan under (Scientific relevance and priorities).

a) Technologies for farmers, households and warehouses to reduce insect menace - thereby enable us to consume nutritionally rich grains

b) Use of such devices in reducing drudgery of insect cleaning (sieving) by women (as experienced in the tribal regions of Adilabad dt. of Andhra Pradesh)

c) Induce creativity among students (school / college) - Educational impact

d) Entrepreneurship development

e) Application of the Indian technologies in food / grain processing industries

6. TECHNOLOGIES DEVELOPED AND THEIR SIGNIFICANCE INCLUDING THE IMPACT ON END USERS (HOME, FARM AND WAREHOUSE)

6.1. TNAU Insect Probe Trap

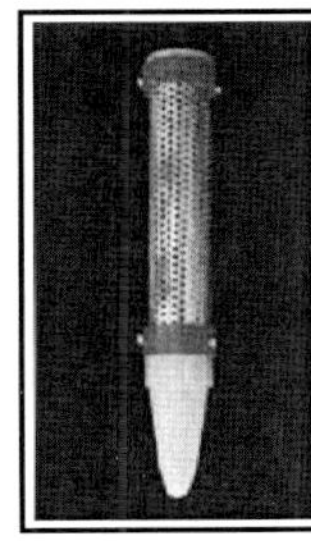

Use of trap is relatively a new method of detecting / trapping insects in stored grains. The basic components of a TNAU probe trap consist of three important parts : A main tube, insect trapping tube and a detachable cone at the bottom. Equispaced perforations of 2 mm diameter are made in the main tube.

Concept

Insects love "AIR" and move towards air. This behaviour of the insect is exploited in this technology.

Method of working

The insect trap has to be kept in the grain like rice, wheat etc., vertically with the white plastic cone downside as shown the figure. The top red cap must be kept to the level of the grain. Insects will move towards air in the main tube and enter through the hole. Once the insect enters the hole it falls down into the detachable white cone at the bottom. Then there is no way to escape and the insects are trapped for ever. The white detachable cone can be unscrewed once in a week and the insects can be destroyed.

Salient Features

No chemicals; no side effects and no maintenance cost.

Significance

- Commercialised during 2002. Around 1,00,000 units have been sold so far across the country. The trap has been included for popularisation under Canadian International Development Agency (CIDA) and Mc Gill University Project on "Food Security in South India" at TNAU, Coimbatore and UAS (University of Agricultural Sciences), Dharwad during 2003-2007.
- CARE (Co-operative for Assistance and Belief Everywhere) world's largest independent, non-profit, non-sectarian, non-governmental, international relief and development organisation with its internal secretariat at Brussels, Belgium is using the TNAU

probe trap technology under integrated child development services in Madhya Pradesh.

- Won Best Creative product award from Chamber of Commerce, South India, 2002.

Recent Achievements

a) A good response from tribal regions of Adilabad district of Andhra Pradesh for the trap. Women are purchasing the trap and using them to control insects in rice / sorghum. They say that the traps are useful in reducing the drudgery of cleaning their grains through frequent sieving.

b) Recently M/s. TATA Chemicals Ltd., Kolkatta has taken up marketing of TNAU probe trap, through TATA Kissan Sanchar in Gorakpur area of Uttar Pradesh after detailed field studies independently by them.

Efficiency

TNAU Insect traps are excellent insect detection devices in food grains and more effective in the detection of stored grain insects namely *Rhyzopertha dominica* (F.), *Sitophilus oryzae* L. and *Tribolium castaneum* (Herbst) in stored food grains both in terms of detection as well as number of insects caught than the standard normal sampling method (by spear sampling). The detection ratio (trap : normal sample) is higher in the trap than that of the normal sampling method by factors ranging from 2 : 1 to 31 : 1. The insects catch is also higher in the probe trap than the normal sampling method by factors ranging from 20 : 1 to 121 : 1.

They are also considered as good mass trapping devices when used at 2 - 3 numbers / 25 kg bin (28 cm dia and 39 cm length). They should be placed at top 6 inches of the grain, where the insect activity is seen during early period of storage. They can remove > 80% of the insects within 10 - 20 days.

6.2 TNAU Pit Fall Trap

Pitfall traps are used for capturing insects active on grain surface and in other layers of grain. (Monitoring and mass trapping tool).

Standard Model

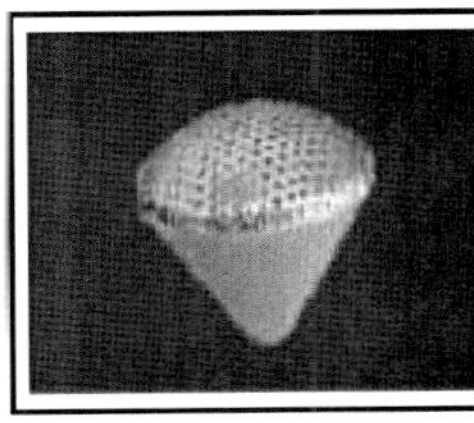

- Standard model of pitfall trap has 2 parts, perforated lid (2 mm (or) 3 mm) and cone shaped bottom portion.
- Application of special coating with sticky material on the inner side of cone to hold trapped insects is necessary
- This procedure is tedious.

TNAU Model

- TNAU model has perforated lid, cone shaped bottom which tapers into a funnel shaped trapping tube.
- Hence sticky coating is dispensed with
- Commercial model is in plastic, simple and economical (cost per trap is Rs. 25/- only).
- Easy to handle.

Significance

- Commercialised very recently and the product was launched on 18th September 2006 by TNAU and available in food grade plastic materials.
- Cost is only Rs. 25/-
- Around 1000 units have been distributed to KVKs for popularisation of the technology
- Because of easy handling and low cost, there is great scope of this model to spread fast across the country.
- No such cost effective and durable device is available so far in any part of the world. Twenty five Rupees investment will be worth to collect plenty of stored grain insects.

6.3 TNAU Two-in-one Model Trap

The probe trap containing the components namely the perforated tube, pitfall mechanism, a collection tube and the cone shaped pitfall trap with a perforated lid and the bottom tapering cone were combined as a single unit. Combination of probe and pitfall increase the trapping

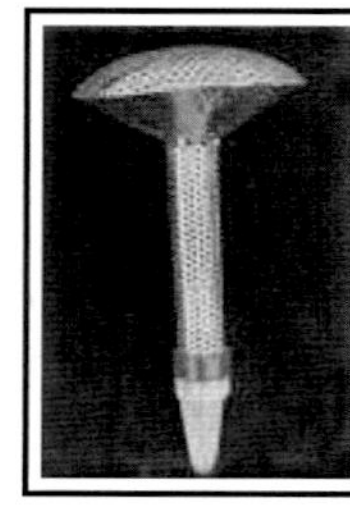

efficiency of insects. Best suited for pulse beetles as they are seen only on grain surface wandering here and there. It does not require tedious procedure like coating the inner surface of pitfall cone with sticky materials before trapping to hold pulse beetles. Beetles are captured alive with this trap, which may facilitate release of pheromone and thereby attract more number of insects.

Significance

i) Commercialised and commercial launching was held on 30.12.2003

ii) Around 100-200 units have been dispatched to UAS, Dharwad for popularisation

iii) This trap technology is one of the promising technologies commercialised under National Agricultural Technology Project funded by World Bank suitable for rainfed eco-system

iv) Spreading fast across the country.

6.4 Indicator Device

It consists of a cone shaped perforated cup (3mm perforation) with a lid at the top. The cup is fixed at the bottom with a container and circular dish, which are to be smeared with sticky material like vaseline.

Farmers, before storing their pulses, should take 200 g of pulses to be stored and put them in the cup. When the field carry over beetles start emerging, due to their wandering behaviour, they enter the perforations and get slipped off and fall into the trapping portions. As they stick on to the sticky materials, farmers can easily locate the beetles and take out the bulk-stored pulses for sun drying. The device with 2mm perforations can be used for cereals, too.

This will help in eliminating the initial population, which acts as the major source for further build up. Thus, timely detection will help the farmers to preserve their valuable pulses during storage. The device is being popularised.

6.5 TNAU Automatic Insect Removal Bin

TNAU insect removal bin can remove the insect automatically. The structure has 4 major parts namely outer container, inner perforated container, collection vessel and the lid. The model exploits wandering behaviour of stored product insects as well as the movement of these insects towards well aerated regions. The grains are held in the specially designed inner perforated container. The space between inner and outer container provides good aeration for the insects. Insects, while wandering, enter the perforation to reach the aerated part and while doing so, get slipped into the collection vessel through a pitfall mechanism provided in the collection vessel. In order to quickly collect the insects, as and when they emerge from the grains, perforated (2 mm) rods are fixed in the inner container.

The container will be useful for storing rice, wheat, broken pulses, coriander etc. The insects such as rice weevil, lesser grain borer, red flour beetle, saw toothed beetle, which are commonly found attacking stored grains can be removed automatically by storing the grains in this container. Within a very short period of 10 days, more than 90 per cent of the insects can be removed from the grains. The container is available in 2 kg, 5 kg, 25 kg, 100 kg and 500 kg capacities.

Efficiency

Grains (paddy and sorghum) stored in Automatic insect removal bin (100 kg and 500 kg) recorded only 1 - 4% damage by insects compared to 33 to 65% damage in the ordinary bin after 10 months of storage. The population of insects (*R. dominica, S. oryzae*) ranged from 0 - 2 / kg in the grains stored in 100 kg Automatic insect removal bin compared to 5 - 191 / kg in ordinary bin after 10 months of storage.

Significance

i. The technology was popularised by the Avinashilingam Deemed University through a project from Department of Science and Technology, Government of India.

ii. Tamil Nadu State Council for Science and Technology sanctioned Rs. 10,000 during 1998 for popularisation of the gadgets among the women folk.

iii. Suitability for developing countries

No granaries can be filled with grains without insects as the harvested produce contain, egg (or) larvae (or) pupae because of 'field carry over infestation' which cannot be avoided in developing countries like India.

So the aim of Storage Entomologist should target this field carryover population (which is generally < 1%) from which adults emerge at 15 to 20 days after harvest in granaries.

If this field carry over population is not cared, then the insect will multiply and contaminate the food grains and complete damage can even be expected within 4 months of storage.

Here comes the role of the automatic insect removal bin which can remove substantial number of emerging adults and sometimes free living larvae also.Significant achievements have happened in popularisation of Automatic Insect Removal bin Technology.

6.6 UV - Light Trap for Grain Storage Godowns

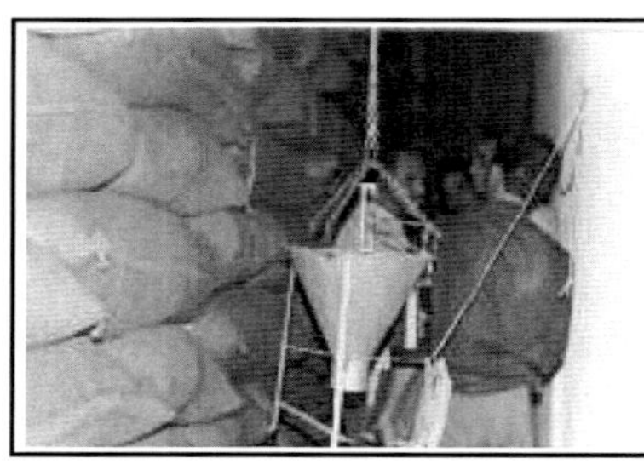

The UV light trap mainly consists of an ultra-violet source (4 W germicidal lamp). The lamp produces ultra-violet rays of peak emission around 250 nano meter. The light is fitted at the centre of a funnel of 310 mm diameter at the top and 35 mm diameter at the bottom. The bottom end of the funnel is attached with a transparent plastic container for collecting the trapped insects. To hang the unit at desired points, three hooks have been provided at the periphery of the funnel. The unit is also provided with a tripod stand.

The UV light trap can be placed in food grain storage godowns at 1.5 m above ground level, preferably in places around warehouse corners, as it has been observed that the insect tends to move towards these places during the evening hours. The trap can be operated during the night hours. The light trap attracts stored product insects of paddy like lesser grain borer, (*Rhizopertha dominica*), red flour beetle, (*Tribolium castaneum)* and saw toothed beetle, (*Oryzaephilus surinamensis)* in large numbers. Psocids which are of great nuisance in the godowns are also attracted in large numbers. Normally 2 numbers of UV light trap per 60 x 20 m (L x B) godown with 5 m height is suggested.

The trap is ideal for use in the godowns meant for long term storage of grains, whenever infested stocks arrive in the godowns and during post fumigation periods to trap the resistant strains and left over insects to prevent build up of the pest populations. In godowns of frequent transactions the trap can be used for monitoring.

Efficiency

It has been found that two traps kept at the corners of the warehouse (60m x 20m x 5m) can catch around 200 insects/day even from a godown where normal sampling did not show any insect presence, thus indicating its effectiveness as a monitoring and mass trapping device. It has been recorded around 3000 *R. dominica* on a single day were collected from single trap kept in a paddy godown.

Significance

National

- The Food Corporation of India has entered into an MOU with TNAU for popularisation of this trap. Already 25 officers have been trained on this line.

- M/S Bannari Amman Sugars Limited, Sathyamangalam utilizes the UV light trap technology under the consultancy project with TNAU for the management of stored grain insects in their godowns.

International

M/s. Madaus Pharmaceuticals Pvt. Ltd Goa, (German Firm) has adopted the UV trap technology under my supervision to detect/mass trap cigarette beetle infestation in their export produce, senna pods/ seeds (*Cassia augustifolia*).

7. TECHNOLOGY OF A NEW EDUCATIONAL KIT AND IT'S IMPACT ON AGRICULTURAL EDUCATION IN INDIA: TNAU STORED GRAIN INSECT PEST MANAGEMENT KIT

Food grains are stored for varying periods to ensure proper and balanced public distribution throughout the year. Among the biotic and abiotic factors which affect the grains / seeds in storage, insect plays a major role in the deterioration of the grains / seeds causing both qualitative and quantitative losses. Often the presence of the insects in store houses is felt only when they hover around, by which time enormous loss and population build up of insects might have occurred. Hence, timely detection of the stored grain insects will help to prevent heavy losses.

TNAU is one of the pioneering institutes in India in the development of detection devices for the stored grain insects. These devices exploit the wandering behaviour of the insects and help in timely detection of them in the stored produce leading to timely control. They include TNAU probe trap, TNAU pit fall trap, Two in one model trap, Indicator device, Automatic insect removal bin and UV-light trap. These devices have been widely used in many places and have received State and National recognitions.

Hence, the Department of Agricultural Entomology, Centre for Plant Protection Studies, TNAU, Coimbatore has developed a "KIT" named as **TNAU-Stored Grain Insect Pest Management Kit** containing prototypes of all the devices along with a CD-Rom about the devices and their use. The kit will be of great use in the popularization of the technologies across the country and will be an ideal "hands - on training" tool for Education, Extension centers (KVK, Plant clinic, save grain centers) and also for private ware houses. This TNAU kit is the first of its kind in the world.

Significance

- The kit has been commercially launched by the TNAU through an MOU with M/s. KSNM Marketing Ltd, Coimbatore.
- Good response from various Agricultural Colleges, KVKs, PHTs Centers across the country. So far around 54 has been sold by the firm.
- For the benefit of the students and staff of agricultural colleges and Home Science Colleges of our country the entire information on above kit is hosted on the website under TNAU-Mc-Gill CIDA Project. www.agrenv.mcgill.ca/agreng/india/files/pest%20trap

- ICAR - Agricultural Engineering Regional Center (ICAE) at Coimbatore has taken up TOT of this technology and spread the technology through Income generation project.
- It is hoped that all the KVKs, SAUs and Home Science Colleges will use this kit for training and education in the days to come.

— xxx —

5

Preservation Problems in Warehouses and their Management

CHADDA, I.C.
Deputy General Manager (Tech /R & D), Central Warehousing Corporation, Hauz Khas, New Delhi – 110 016.

SUMMARY

The paper throws light on the national scenario of preservation problems in the warehouses and series of management plans for insect problems in the warehouses.

Key words : Preservation, Warehouse, Management.

1. INTRODUCTION TO WAREHOUSING

Warehousing is closely linked with the development of agriculture, trade and industry and no economic activity is free from warehousing. The idea of Public Warehousing in India, took shape mainly because of widespread rural indebtedness. It was felt that overwhelming majority of the farmers were heavily indebted primarily because they were forced to sell their produce immediately after the harvest at very low prices, just to repay the loan obtained before harvest or loan/ interest running outstanding generation to generation.

In 1928, the Royal Commission on Agriculture while pointing out serious deficiencies in the Agricultural Marketing had first recommended the establishment of licensed warehouses. Thereafter, various commissions and committees had gone into finding ways to reduce rural indebtedness. It was observed that the volume of finance supplied by the Commercial Banks to the farmers could be considerably increased by improving the arrangements for marketing of crops by grading, standardization and creation of proper storage facilities. Keeping this aspect in view, the need for establishment of warehouses in which the goods could be stored against a negotiable receipt, for obtaining loans from the banks was felt.

Food security for the entire population coupled with better nutrition, sustainability of agricultural development are our national priorities which can be taken care of only if good infrastructures warehousing facilities are available, to store surplus food grains to meet out the food requirement of our entire population. The need of providing modern efficient system of warehousing to ensure maximum returns to the growers with better opportunities, for agricultural produce and minimizing the wastages at this stage is, but felt that CWC along with its seventeen subsidiaries in the shape of State Warehousing Corporation with such a mandate has been in a position to provide warehousing and logistics services to its clientele.

Central Warehousing Corporation (CWC) came into existence on 2-3-1957 in pursuance to the Agricultural Produce (Development & Warehousing) Corporation's Act, 1956 later repealed and replaced by Warehousing Corporations Act, 1962 with a meagre capacity of 7000 MTs in hired godowns at 7 centres. It has steadily increased its warehousing capacity and Is now operating 514 Warehouses with a total capacity of 10.26 million tonnes. CWC handles variety of commodities of agricultural origin, farm inputs, industrial goods etc.

Ever since its inception, the Corporation has been giving highest priority to storage of agricultural produce and inputs at a very reasonable cost. In order to encourage the farming community for using public warehousing facilities, CWC offers a rebate of 30% in its storage charges for the stocks deposited by farmers.

2. PRESENT SYSTEM OF STORAGE

Presently bagged foodgrains are carried from the mandis to the storage points and there is no system of bulk handling in India. In order to ensure all around supply, there is a need of storage at appropriate level. Storage is basically done by farmers, traders and the Govt. agencies. Nearly 70% of the production is retained at the farm level in India, which is stored using the traditional methods of storage. This is a place where considerable qualitative and quantitative losses take place.

The storage capacity available with major agencies is given in (Table 1).The main public agencies involved in the storage of foodgrains are Food Corporation of India (FCI), Central Warehousing Corporation (CWC) and 17 State Warehousing Corporations (SWCs).

Table 1 : Storage capacity available with major storage agencies as on 1.2.2007

Agency	Covered		CAP		Grand Total
	Owned	Hired	Total	Owned & Hired	
FCI	12.94	8.90	21.84	3.07	24.91
CWC	6.70	2.16	8.86	1.45	10.31
SWCs	11.89	6.42	18.31	0.72	19.03

2.1 Conventional Godowns

The foodgrains in the bags are stacked by leaving a proper alleways for inspection and movement of the foodgrains. These covered godowns are constructed scientifically based on the design given by BIS. Scientific storage, techniques are adopted for control of the insects/ rodents and other pests. The stocks are regularly inspected for determining the intensity of insect infestation as well as categorization. The principle of First In First Out (F.I.F.O.) needs to be practiced to avoid the prolonged storage of foodgrains in the godowns. The storage system presently being adopted is generally the covered godowns. However, at a time of scarcity of the storage space, the stocks are even stored in the open.

2.2 Open Storage System (CAP)

The open storage of foodgrains is done under covered and plinth system (CAP). In this system foodgrains (wheat and paddy only) are stored in open complex. The grain bags are stored on the wooden crates on the plinths made of brick and cement. The foodgrains are covered with specially fabricated polythene covers and tied with nylon nets and ropes. This is a temporary storage system and the stocks are likely to get damaged if allowed to continue for a longer period, due to high velocity winds, rains, rats, birds, etc. This system calls for constant vigilance and regular monitoring of quality of the stocks.

Problems in the existing system of storage and transportation

The existing system of handling the food grains is having many shortcomings. Some of them are given below:-

- Slow, inefficient handling of the grains in the mandis, having no facility for mechanisation.
- Poor utilisation of the market yards as the operations are labour intensive leading to lesser labour productivity.
- Qualitative and quantitative grain losses/damages in all the stages of procurement, storage and transportation.

The present manual operations are not very effective in the peak marketing seasons to cope up with the increased volume of business. Thus, there is a need for introducing mechanisation in the existing system of handling storage and transportation in a phased manner. National Storage Policy (2000) announced by the Government of India is a right step in implementation of bulk storage handling and transportation of foodgrains.

3. MANAGEMENT PLAN FOR INSECT CONTROL IN WAREHOUSES

The grains to be stored in the godowns require very careful handling to prevent any sort of quality deterioration. The good quality grain i.e. the grain with minimum foreign matter, having moisture content within safe storage limit is less likely to suffer insect attack than the poor quality grain. The good house keeping inside and outside the godown premises, is and essential part of the insect pest management. This not only helps in reducing the pest activity but also improving the insecticide application. There is necessity for regular inspection of the stocks for monitoring of the pest population, its identification and carry out prescribed prophylatic and curative treatments. The stacking in the godown requires to be undertaken in a scientific manner, keeping sufficient alleyways for inspection and carrying out chemical treatments, wherever necessary. The regular monitoring of the pest population helps in planning proper pest control measures.

3.1. Fumigation

Fumigation plays a key role in control and management of infestation in commodities stored in tropical countries like India. It has proven to be versatile technique and great reliance has been placed on it to underpin the success of other pest and quality control measures, particularly in countries like India where admixture of grain protectants is prohibited. It is a curative treatment which adds to the storage hygiene by keeping infestation free storage environment to comply with regulatory requirement.

Methyl Bromide (MBr.) and Phosphine are the most commonly used fumigants. Inspite of having some advantages, MBr. as an fumigant is on its way out, as it has been identified as a serious ozone depleting agent with a Ozone Depletion Potential (ODP) of 0.7. The Clean Air Act in the US declares that any product with an ODP of 0.2 or greater will be eliminated. The Montreal protocols 120 plus signatory countries have also endorsed the same. This makes phosphine as the most accepted fumigant for stored products. There are several developments in our knowledge of phosphine which lend positive support to the continuous use of phosphine, as a fumigant (Banks 1994). The major threats to phosphine use are occupational health, safety and environmental risks and insect resistance (Darby and Annis 2003). Meaklim (1998) concluded that the main threats to continue phosphine availability were accidents and usage by untrained people.

3.1.1. Fumigation with phosphine

Phosphine is generated by the action of moisture in the air on the tablets, pellets, sachets or plates of aluminum magnesium or zinc phosphide. The tablets generally weigh 3 grams and release one gram of Phosphine whilst the pellets weigh 0.6 grams and release 0.2 grams of Phosphine. Individual sachets weigh 34 grams and release 11 grams of Phosphine. Phosphine blankets are also used in certain countries like Australia which are available in a single or multiple plate pack in the form of strip. The main advantage of the blanket pack is that these can be removed from the grain prior to out loading thus leaving no physical residue of the spent material in the grain.

In India, the tablets are distributed on the stack after keeping them in paper / enamel plates and covering the stack by a gas proof cover for a period of about 7 days. Earlier naked phosphine tablets were being distributed directly on the bags which created a lot of problems in removal of the residue. The powdery residue in the form of alumina though not toxic was a source of big irritation to the godown operators. At times, this residue also contained, small amount of undecomposed phosphine, thus causing health and safety problems. The left over residue with the application of prophylactic treatment formed a thick paste thereby aggravating further problems in aeration and penetration of the fumigants.

3.1.2. Fumigation with mixture of phosphine and carbon dioxide

Increased carbon dioxide accelerates respiration of the insect thereby making them more susceptible to phosphine. Mueller (1994) reported that insects are stressed by the increased level of carbon dioxide and heat, which allows lower level of phosphine to be more effective in achieving 100% mortality of all live stages including the egg stage in shorter periods of time. The insect respiration can be increased by 50% by increasing carbon dioxide level to 3%. The respiration gets increased to 300% when carbon dioxide level are raised to 5% (Mueller). The mixture of phosphine and carbon dioxide has the capacity to reduce the resistance level to phosphine in population of *Rhizopertha dominica* and *Sitophilus oryzae* (Athie *et.al.*, 1998).

The trials on mixture of carbon dioxide and phosphine carried out by a team consisting of CWC, Indian Grain Management Research Institute, Hyderabad and Excel Industries using the phosphine generator made by this team has established that 1.3 gm. of phosphine and 35 gm. of carbon dioxide per cubic mtr. is sufficient to kill all the test populations including *Trogoderma granarium*. The dosage of phosphine could be reduced to 1 gm. if, population of *Trogoderma* was not present (Chadda *et al.,* 2004).

3.1.3 Fumigation failures

If live insects are found after fumigation, the problems could be caused by :-

- The fumigant was not retained for long enough.
- Not enough fumigant was added.
- Insects entered the storage after fumigation.
- The insects were resistant to the treatment.
- Lack of monitoring devices.

In most cases phosphine fumigation failures are caused by the fumigant not been retained for long enough. In bagged cargo, fumigations as is being done in India, the type of fumigation sheets selected for fumigation plays a very vital role in success or failure of the fumigation. The black polythene covers even used by many public sector agencies often develop pin holes in a short period leading to leakage of the phosphine and ultimately ineffective fumigation. Fumigation requires application of the correct dosages and its proper distribution to ensure mortality of the target insects. In the fumigation process, the fumigation sheets are generally sealed using single layer of sand snakes, which is actually not sufficient to make the cover air tight. More cases of fumigation failure are noticed, when room fumigations are carried out, which permit leakage of the chemicals. This results in not only wastage of the chemical but provision of sub lethal dosage which can become the cause for development of the resistance. Monitoring of the gas concentration during fumigation is generally not being done in India. In absence of the monitoring, it is really not possible to find out the causes of fumigation failure.

3.1.4. Phosphine resistance

Phosphine is being used for many years throughout the world for the control of grain pests. Strong resistance to phosphine in primary pests of cereals were identified in India (Rajendran and Narasimhan, 1994). Many times fumigations are reported to be failures but, it needs to be properly investigated whether failure of fumigation is because of improper practices of fumigation or because of insect resistance. It is an established fact that because of low toxicity or because of residue problem, fumigant may become ineffective. Dosages of the fumigant and exposure period are extremely important. There has to be a continuous system of monitoring for fumigation resistance so that if it develops, it is detected quickly and early warning system is in place to enable a strong responsibility when commercially damaging resistance appear. The causes of insect resistance needs to be tackled by identification of strategies for implementation of best practices of phosphine. It is all the more necessary to extend the useful life of phosphine, keeping in view the enormous cost of around 60 million US $ for development and registration of new pesticides.

3.1.5. Alternative fumigants

There is no immediate feasible alternative to phosphine but keeping in view the development of resistance strains in some cases, it is necessary to work on alternative fumigants. The following list though not exhaustive it covers the most likely current or up coming options.

Eco_2 Fume

It is non flammable liquid phosphine fumigant gas mixture (2% phosphine with 98% of carbon dioxide) to maximize phosphine's superior fumigation properties by eliminating its spontaneous flammability hazards. This is more user friendly leaving no residues and capable of eliminating all insect stages.

Mixture of Phosphine and Carbon Dioxide

Considering the advantages of carbon dioxide phosphine gas generated through a phosphine generator, if mixed with suitable combination of carbon dioxide can prove to be highly successful.

Sulfuryl Fluoride (Profume)

It is a rapid fumigant which is already registered by Environmental Protection Agency (EPA) of USA as a grain fumigant in many states of USA and also accepted by Australian Quarantine Authorities (AQIS). It is a hazardous gas or liquid under pressure and is being considered as a replacement to MBr.

Ethyl Formate

Ethyl Formate has a long history for use as a fumigant for stored products in general and is currently registered as fumigant for dried fruit treatment in Australia. It is a liquid fumigant and kills the insect quickly and its residues break down rapidly to non- poisonous and naturally occurring products i.e. formic acid and ethanol.

Vapormate

CSIRO Australia has now recommended vapormat (6.7% by weight mixture of Ethyl Format in carbon dioxide). It is considered advantageous due to fast kill of insects, fast breakdown of the residues to natural products. It helps to enhance insect activity and is effective against variety of storage pests. Exposure period between 3 to 4 hours gives mortality of the storage pests.

Propylene Oxide

This is colourless liquid with sweet ether like odour. This is extremely flammable liquid. This is registered in USA and is the potential alternative to MBr. for rapid disinfestations of the commodity. At 30^0 C, 4 hours exposure is required for killing most of the pests. As per the EPA, 8% propylene oxide mixed with 92% carbon dioxide is permitted for fumigation process at a temperature of 125^0 F for a maximum of 48 hours.

Carbonyl Sulphide

Carbonyl Sulphide gas can become an effective and safe alternative to phosphine. This has got a significant potential as a new fumigant for use as an alternative to phosphine. This comes as a gas in pressurized cylinders and can be applied quickly and easily. It is reported to be effective against all stages of most stored grain insect pests. It works out more rapidly than phosphine with treatment time of about 2 to 5 days compared with 5 to 10 days for phosphine.

4. ROLE OF NEEM IN INTEGRATED PEST MANAGEMENT

Pest control potential of neem has been identified as environmentally soft natural pesticide. This is very safe since it does not contain any toxic residue. It is harmless to the natural predators, parasites and no resistance to pests has been noticed and these products are totally biodegradable, having no ecological or environmental problems. Various studies have confirmed effectiveness of neem products against storage pests like *Rhizopertha dominica, Sitophilous oryzae, Oryzaephilus* sp. and *Corcyra* sp., either alone or in combination with grain fumigation. While neem treatments cannot replace completely, the chemical pesticides, the amount of pesticides could definitely be reduced thereby decreasing the pesticide load in foodgrains. Thus, neem products can be ideally suited for integrated pest management. CWC is also working on the field trials using 'Azadirachtin' 1500 ppm developed by M/s I.T.C. Ltd., Guntur as grain protectants.

5. EMERGING ISSUES

The grain industry in India is passing through a crucial period, in which efforts are required not only to maintain the quality of the foodgrains but also to increase the quantity. Regular recurrence of pest resistance due to adoption of present control system or due to improper fumigation practices could pose a serious threat in the field of pest control. The general problems requiring serious attention are listed below

Phosphine is the only fumigant relied by the grain industry for maintaining insect free grains. The failures in the fumigation process do occur due to various reasons. The main reason appears to be leakage of the fumigant due to non availability of gas tight environment The adoption of high level of gas tightness, using proper sealing methodology is the only answer to have effective fumigation which can help in reduction of dosages besides creating environmental friendly atmosphere. Thus, there is a need to work out better methodology for sealing of cargo under existing system of storage.

It is an accepted truth that fumigation operators in India are not giving any cognizance to monitoring of the gas concentration during fumigation process. The importance of regular monitoring in ensuring effective and efficient fumigation cannot be ignored. It would not be wrong, if it is said that "NO MONITORING MEANS NO FUMIGATION". Thus there is a necessity for stringent monitoring of the gas concentration and also capture of the residual gas. Latest technology in this regard is available, but support from industry and government is necessary.

The development of alternative fumigants is becoming a matter of importance because of threats, to long term use of only two widely used fumigants Methyl Bromide and Phosphine. Methyl Bromide is in the process of being phased out, due to changing social and environmental standards. The development of resistance to certain pests against phosphine has necessiated search for alternatives to replace this rapid action fumigant. The development of new fumigant is not easy. The acceptable grain fumigant must not damage the grain, must be relatively cheap, environmental free, convenient to apply, safe to handle and capable of killing all stages of insect pest in shorter duration.

The dosages of the pesticides for Prophylatic and Curative treatment are fixed irrespective of the type of pest population and climatic conditions. In order to have effective treatment, dosages need to be worked out selectively. This will not only help in minimizing the usage of pesticides but more environmental friendly approach to solve pest problems.

The fumigation system at the farm level is still in very very primitive stage where farmers feel satisfied about success of the fumigation after putting the tablets into the bin/gunny bags etc. There is no methodology to find out success or failure of the fumigation process. In this regard, introduction of "**PHOS- CARD**" is suggested. The Phos Card is a simple card of the size of credit card, having copper strip embossed on it. Since phosphine reacts with copper, the activity in terms of shade of copper strip can be correlated to approximate concentration of phosphine. This could very well be introduced for proper monitoring of the fumigations at the farm level.

Keeping in view the resistance of certain grain protectants to certain pests, in some countries like Australia, mixture of grain protectants are recommended because no single protectants will give long term protection against all resistance species. The problem is universal but no such mixing of the different grain protectants is presently allowed in India. This aspect needs to be examined by working out various combinations. Such feasibility will ultimately help in extending the useful life of the pesticides.

Pest resistance to pesticides is a universal phenomenon. It, therefore, calls for identification of the suitable strategy to counter this problem.

The controlled atmosphere system protects the grain quality, controls insect and mite pests. This refers to altering the normal grain atmosphere by introducing inert gases in a sealed storage. The combination of inert gases with different fumigants needs to be tested with the objective to reduce the dosages of the fumigants and to avoid various problems related to with excessive use of pesticides.

The neem products has been identified as environmentally soft natural pesticides. The effectiveness of this product alone are in combination with existing grain protectants / fumigants needs to be evaluated.

Food represents the very essence power of the creation and stands at the cross road between the infinite force of nature and the biological wonder of the land. It is true that the pest create havoc by substantially damaging the foodgrains but continuous application of pesticides is very risky. It, therefore, calls for development of strategy, which needs the highest standards of pest control in consumption with good management practices. The adoption of integrated pest management system for prevention of the spread of pest population by controlling natural breeding conditions followed by measured use of either pesticides or organic pesticide or biological application.

6. CONCLUSION

In order to provide insect free and quality food to mankind, it is necessary that researchers in the field of food storage must join hands with the commercial players like Warehousing agencies and pesticide industry to work out alternative pest control strategies, for safeguarding interests of the food grain industry in India having highest standards but involving lesser usage of pesticides, in association with good management practices.

References

Athie et al. 1998. Effects of carbon dioxide and phosphine mixtures on resistant populations of stored grain insects. *J. Stored Prod. Res.*, 34:1, 27-32.

Banks H.J. 1994. Fumigation - an endangered technology pp 2- 6 Editors E. Highley *et. al.* In Proceedings of the 6th International Working Conference on Stored Product Protection, Vol. I, CAB International, Canberra.

Chadda I.C., Jayraj. K. and Dhuri A.V. 2004. New method of Phosphine and Carbon Dioxide application and its optimization. Proceedings of the International Conference on Controlled Atmosphere & Fumigation in Stored Products 8th to 13th August 2004 at Gold Coast, Australia.

Darby, James and Annis Peter C. 2003. Integrating fumigation and aeration *In* pp 126- 133 Proceedings of the Australian Post harvest Technical Conference, Canberra 25-27th July, 2003.

Meaklim, J. 1998. Phosphine toxicity; are phosphine users, or the general community, at risk of adverse health affects. In: pp 119-125 Banks, H.J., Wright, E.J. and Damcevski, K.A., (Eds.). Stored grain in Australia. Proceedings of the Australian Post harvest Technical Conference, Canberra, 26-29 May 1998. Canberra, CSIRO Stored Grain Research Laboratory,.

Mueller D.K. 1994 A new method of using low levels of phosphine in combination with heat and carbon dioxide, *In* pp Editors E. Highley et al, Proceedings of the 6th International Working Conference on Stored product Protection, Vol. I, CAB International Canberra. 123-125

Rajendran, S. and Narasimhan, K.S. 1994. The current status of phosphine fumigations in India. Wallingford, Oxon, UK, *In*: pp 148-152. Highley, E., Wright, E.J., Banks, H.J. and Champ, B.R., ed. Stored product protection. Proceedings of the 6th International Working Conference on Stored–Product Protection, Canberra, Australia, April 1994 CAB International.

— xxx —

6

Food Corporation of India : A Role Model Organization in Pest - Free Food Grain Storage

ARUNACHALAM, A.S.
Assistant General Manager (QC), Food Corporation of India, Regional Office, Chennai – 600 006, Tamilnadu.

SUMMARY

Emphasis the paper focusses the need and scope for insect free grains in storage while giving on use of certain insecticides and bio pesticides. Final conclusion indicates role of the integrated pest management approach in storage of grains.

Key words : FCI, Grain Storage, Support Price

1. INTRODUCTION

Food Corporation of India (FCI), the largest single national food grains handling Public Sector Organization was set up by enactment of Food Corporation Act 1964 with the primary objective of producers welfare and consumers protection in India. Gigantic procurement of food grains under Support Prices from the producers, storing and preserving both operational and buffer stocks and distribute with same quality to vulnerable sections of the society at reasonable prices, are some of its prime functions.

Preservation is one of the main functions of FCI to retain the basic characters the grains during storage so as to ensure the supply and satisfy the basic nutrients and to provide energy to the consumers.

2. FOOD CORPORATION OF INDIA AND GRAIN STORAGE

In FCI, foodgrains are stored in scientific godowns which have been built on the raised platform to avoid entry of rodents, smooth finish to walls and avoid sheltering of insects. Food grains are being stacked on dunnage materials like Polypallets, LDPE, GFM sandwiched between two layers of mats. These dunnage materials are ensured to give protection to the bottom to avoid moisture migration.

3. TREATMENT OF FOOD-GRAINS

Prophylatic treatment of grains controls the surface, crawling and cross infestation it is given once in a fortnight with Malathion or once in three months with deltamethrin is used. Food grain bags are sprayed on the surface by these two contact insecticides to arrest the activity of the insects.

In addition, air charging of empty space is also undertaken by spraying with DDVP once in a week to control the crawling and empty surface infestation. Such chemicals are not sprayed on either foodgrain bags or naked grains.

Curative treatment is undertaken to control the hidden and resident infestation pests. Fumigation of the foodgrains with aluminium phosphide is being carried out in FCI. The aluminium phosphide acts by release of phosphine gas along with carbon dioxide and ammonia which are also lethal to the pest population besides controlling the inflammability of the phosphine gas.

For fumigation of the food-grains, the rubberized polyethylene covers and or new LDPE polythene covers and multilayered cross laminated fumigation covers are being used in the godowns of FCI. The stacks are made air - tight by mud plastering or sand snakes to retain the maximum gas. The dosage of the chemicals and their exposure time are detailed below for elimination of the pests.

Type of Treatments	**Name of chemicals**	**Formulation**	**Dilution/ Dosage**	**Periodicity of treatment**
Prophylatic	Malathion	50 EC	1:100	15 days and 21 days in winter
-do-	DDVP	76 EC	1:150	As and when required
-do-	Deltamethrin	2.5 MM	40g/lt	Once in 3 months
Curative	Aluminum/ Phosphide	_	5 Tablets / MT	Pre/ Post monsoon and whenever required

— xxx —

7

Significance of Grading and Preservation of Commodities in Storage

KANNAN, V.
Regional Manager, Central Warehousing Corporation, Regional Office, Saidapet, Chennai - 600 015, Tamil Nadu.

SUMMARY

The paper underlines the value of proper grading of commodities and preservation techniques for sustained keep up of quality of the food commodities as practiced in the CWC.

Key words: Grading, Preservation, CWC.

1. INTRODUCTION

Grading and preservation of kinds of commodities are important activities in the grain storage. The main objectives of the Agricultural Produce (Development & Warehousing) Corporation Act, 1956 (Later on repealed by Central Warehousing Corporation Act, 1962) are:

- Creation of a negotiable paper to provide an instrument for extension of credit through commercial banks for the benefit of producers, dealers, etc
- Assistance in orderly marketing by introduction of standard, grade,specification and warehouse receipts.

- Grading of agricultural produces is an important aspect of warehousing. So far, grade specifications have been evolved for more than 120 agricultural commodities and Code of Storage Practice for more than 300 -commodities, including chemicals and notified commodities.

2. GRADING OF COMMODITIES

Grading indicates the commercial value and health of the stock. It is an easily understandable trade language. It also gives an assurance to the buyer of the quality characteristics of the commodity he has placed orders. The seller is also quality conscious as graded stocks fetch a premium over non-graded stocks. It helps in obtaining orderly marketing and advance of loans by banks.

The Directorate of Marketing & Inspection, by forming an expert committee, has evolved national grade specifications for almost all the agricultural commodities including processed food. These grade specifications are adopted at the warehouses. Normally, the grades are mentioned as I, II, III or IV. Grading is done based on % of various refractions. Such refractions are foreign matter, damaged and weevilled grains, broken, immature / shrivelled and moisture content. As grading is an important aspect of warehousing activity, all the warehouses are provided with a laboratory and the required apparatus and equipment for grading. When the stocks are received for storage, representative sample will be drawn, analyzed and the correct grade mentioned in the warehouse receipt.

3. STORAGE AND PRESERVATION AT WAREHOUSES

Warehousing Corporation as per the terms and conditions (available on the reverse side of the warehouse receipt) assures to maintain quality and quantity and to return the stocks in the same condition. In order to fulfill these obligations, consignment of stocks received is carefully preserved during storage. Weighment of the stock is also taken both at the time of receipt and issue. Standard bags (variation between weights of two bags shall not be more than 0.5% of the declared standard weight) are weighed 10% while non-standard packages are weighed 100%.

3.1. Storage Practice

The space meant for the storage is determined based on commodity factor i.e. characteristic of the commodity and capacity factory i.e physical dimension of storage area and load bearing capacity of floor (3.55 MT per sq.mtr). The stocks are accepted on stacks which are rectangular blocks. Out of the total carpet area, normally 73% is used for storage leaving 27% vacant area for inspection and operational convenience. Stocks are accepted over a dunnage material, which prevents damage to bottom layer bags due to seepage. The stack height is restricted depending upon the type of the commodity and package. The stocks are also insured against various contingencies like fire, flood, theft, burglary etc.

3.2 Preservation of Stocks

In order to maintain quality and quantity during storage, a specific drill is prescribed. The code of storage practice is scrupulously followed for each commodity Fortnightly inspection of the stock is conducted by qualified and trained technical staff and the stocks classified and categorized. Fortnightly inspection of the stocks helps to

- monitor condition of the stock in storage
- plan and execute prophylactic and curative treatment
- prevent deterioration of quality of the stock.
- determine priority for release of the stock

4. SIGNIFICANCE OF GRADING, PRESERVATION IN FUTURES TRADING

CWC, with its expertise in grading and preservation of stocks has been able to play a very vital role in Online Futures Trading of Commodities. The CWC could earn the faith of the buyers who purchase the stocks without even seeing them, let alone verifying the quality.

Independent third party grade certification has helped the futures trading in commodities, like black pepper, rubber and cardamom to achieve impressive results.

To conclude, the CWC serves the common man in the storage of grain and other agricultural commodities at its premises.

— xxx —

8

Biology and Infestation Behaviour of Lesser Grain Borer, *Rhyzopertha dominica* Fab. (Coleoptera: Bostrychidae) a New Storage Pest on Finger Millet, *Eleusine coracana* (L.) Gaertn

GURURAJ AINAPUR* P. S. JAGADISH[$], K.S. JAGADISH AND S. SUBRAMANIYAN[#]**

*Department of Entomology, UAS, GKVK, Bangalore-560065;
[$]Professor and Technical Project Leader, All India Co-ordinated Small Millet Improvement Project (AICSMIP), UAS, GKVK, Bangalore-560065;
**Entomologist, AICRP on Sunflower, UAS, GKVK, Bangalore-560065, and
[#]Professor and Storage Entomologist, HPT Scheme, UAS, GKVK, Bangalore-560065.

SUMMARY

Investigation on the biology and feeding behaviour of R. *dominica* on finger millet revealed that larvae underwent four instars and the total developmental period of the immature stages ranged from 40 to 51 days with an average of 44.69 days. The larva converted the grains into cluster while feeding, by using its excreta and frass. Pupation occurred inside the grain cluster. The nature of the larval growth was unique on ragi grains as compared to other cereals.

Key words: Fecundity, Finger Millet, Infestation Behaviour.

1. INTRODUCTION

Finger millet is an important millet crop of India. Southern India, particularly Karnataka is a major producer of finger millet which is the staple food crop of this region. By and large finger millet in storage is known to be free from insect pests and diseases. This myth was recently broken when a sample of finger millet was brought to the notice of the Entomologist at the All India Co-ordinated Small Millet Improvement Project, University of Agricultural Sciences, Bangalore. On examination, it was found that the ragi grains were heavily infested by stored pest identified as lesser grain borer (*Rhizopertha dominica* Fab.), which is known to infest mainly wheat, sorghum, paddy, maize and bajra. *R. dominica* was rated as one of the major insect pests of stored cereals along with *Sitophilus oryzae* (Linnaeus) and *Sitotroga cerealella* (Oliver), in the tropics. So far, *R. dominica* has not been reported on finger millet and necessitated this study to track biology of *R. dominica* on finger millet, its nature of damage and symptoms.

2. MATERIALS AND METHODS

Stock culture of *R. dominica* was used from freshly emerged adult beetles obtained from Central Food Technological and Research Institute (CFTRI), Mysore. They were maintained on finger millet of 12 per cent moisture content, kept in plastic jars covered with muslin cloth. The dead beetles were removed periodically. Life history studies were conducted on grains of finger millet under laboratory conditions during the period January - May 2004 when the laboratory temperature and relative humidity were in the range of 24 to 29 ^{o}C and 68 to 75 per cent respectively.

2.1. Egg Collection

The method for collecting *R. dominica* eggs is a modification of that used for collecting *Oryzaephilus surinamensis* devised by Elek, (1994). Cultures were used for egg collection after 6-8 weeks at laboratory conditions. A stack of at least three filter papers (5 cm diameter) was buried just under the surface of the culture and left undisturbed for about 48 hours. Each filter paper was first slightly bent so that layers in the stack were not completely flat. The eggs were laid in rafts of two to six eggs or clusters of several adjacent rafts both on the outer surfaces and around the edge in between the layers of paper. Since the eggs were firmly stuck onto the paper, tapping each filter paper gently when the stack was removed from the culture dislodged flour and adult beetles. Eggs laid in between the layers were cleaned of frass and flour. The stacks of the filter papers with eggs were placed in petri dishes until they hatched. From this the incubation period was tracked.

The gap between the layers of paper was critical. If too close the eggs were stuck onto both bottom and top layers and chorion was damaged when the layers were separated, but if not close enough the adults were not attracted to lay eggs. The layers of filter paper can be held together with small pieces of tape or a staple in the centre so that they are not separated by activity of adults, particularly if small pieces of paper are used for collection. The white eggs were easier to see on dark-coloured or black paper. These eggs were used to study the biology.

2.2. Biology

Eggs were obtained by the procedure as explained earlier. On hatching, the first instar larvae were carefully released into the tubes (7.5 X 2.5 cm) with the help of fine camel hair brush, having broken grains of finger millet.

The larvae formed a cluster of grain around them and started feeding inside the cluster. Therefore, everyday three grain clusters were dissected and observed carefully to note the stage of the larva until the completion of developmental period, commencing from release. Sufficient number of tubes were maintained till the end of the observation.

Presence of moulted skin and head capsule inside the grain clump was made use of for determining the larval instars. Measurements of the size of head capsule, thorax and body length were also recorded, using a suitably calibrated ocular micrometer mounted on a binocular microscope. The dissection of grain clump was done till only pupal stages were seen. The emergence of the adult was observed from the same stock and pupal period too was determined.

2.3. Ovipositional Studies

Newly emerged adult beetles obtained from the culture were sexed based on the differences in the colour on the ventral side of the fifth abdominal segments (Stemley and Wilbur, 1966).

A pair of beetles along with one gram of finger millet grains was enclosed in a glass vial, and the open end covered by cotton plug.

Observations on mating behaviour, pre-ovipositional and ovipositional period were made. Number of eggs laid per female per day was recorded till egg laying ceased or till the death of the female beetle. In all, 15 pairs were enclosed separately in 15 tubes.

2.4. Adult longevity

The ability of the adults of *R. dominica* to live in the presence and absence of food was determined by enclosing male and female adult beetles obtained from culture separately. Fifteen such vials were maintained for each of the males and females without food and similarly another set of 30 vials, 15 for males and 15 for females, each were set up with finger millet as the food. The temperature and relative humidity prevailing in the room was recorded and it ranged from 24 to 28 °C and 68 to 75 per cent, respectively.

2.5. Sex Ratio

Sex ratio among the adults of *R. dominica* obtained from the culture maintained in the laboratory as well as from those occurring naturally in the godowns was determined.

In the case of laboratory maintained culture, 25 adults were randomly picked and sexed based on the external colour differences on the ventral side of the 5th abdominal segment as given by Stemley and Wilbur (1966).

Similarly, 25 adults collected from warehouses were sexed and sex ratio determined. The sex ratio was worked out based on the number of females to males.

2.6. Nature of Damage and Infestation Behaviour

Feeding behaviour

The first instar larva of *R. dominica* usually bore into the grain and completes its development in case of cereals. In finger millet even though grain size is small, the infestation by *R. dominica* could be seen. Constant observation was made on feeding nature, growth and development of *R. dominica* on finger millet.

Amount of food required by individual larva

Twenty-five tubes filled with 1g of broken finger millet grains of moisture content 12.35 per cent (measured using hot air oven method) were used. Two such sets were maintained throughout the study period. These sets were used to determine the amount of food required by single larva.

Observation on the amount of food required for individual larva was made by subtracting final weight from initial weight.

3. RESULTS AND DISCUSSION

The beetles copulated under laboratory conditions during day time when there was bright light. The mating period ranged from 50 to 123 min (mean of 82.5 min). Repeated mating between a pair was not observed. The pre-oviposition and oviposition period ranged from 6 to 9 and 28 to 45 days, with an average of 7.4 and 34.2 days, respectively (Table 3).

The eggs were generally laid loosely and also in groups of 2 to 14 in the frass (mixture of flour and faecal matter) produced, in some cases glued to grains and were also found in the cavities of broken grain.

3.1. Biology of *R. Dominica*

Freshly laid egg was glistening white in colour and gradually became rosy and opaque with wrinkles appearing on the surface. Being cylindrical in shape, it was gently rounded at one end and somewhat pointed at the pedicel end. Observations pertaining to egg measurement are 0.62 to 0.66 mm length; mean of 0.63 mm and diameter 0.20 to 0.22 mm; mean 0.21mm (Table 2) which are in close agreement with those of Potter (1935).

Measurements of Egg

Length: 0.62 to 0.66 mm (mean: 0.63 mm)

Width: 0.20 to 0.22 mm (mean: 0.21 mm)

The first instar larva was in a curled position in egg shell. In the present study, prior to hatching, the chorion became rather transparent through which the first instar larvae could be seen lying in a curled up position. The observation relating to egg period was 5 to 9 days (mean of 7.09 days) on finger millet but the same was 6.3 days at 30 °C in wheat (Elek, 1994). The minor differences observed in the present investigation could be due to the variation in the host used as well as in the geographic locations.

Freshly emerged larva was white with a brownish head. The hairs on the abdomen are longer than those on the thorax and head. The last abdominal segment has a spine curved and directed posterior on the dorsal surface. Four ocelli are present on each side of the head. The first instar was active and moving in a looping fashion. Second instar larva was seen with the head slightly retracted into the prothorax. In third and fourth instars, the head was found completely retracted into the prothorax. The movement of the larva was restricted at this stage and both the third and fourth instar larvae were lying in curled position. The abdominal segments 5 to 8 were very much extended in the fourth instar. The present descriptions are of the four larval instars are almost in agreement with the findings of Potter (1935). The observations pertaining to the fourth instar larval dimensions (2.9 mm) (Table 2) is in close agreement with that of (2.8 mm) Potter (1935).

The period occupied for larval development was 6 to 7, 7 to 9, 7 to 8, and 8 to 10 days with a mean of 6.74, 7.82, 7.04 and 9.27 days, respectively for four larval instars under laboratory conditions (Table 1). Elek (1994) revealed that it was 9.6, 5.7, 4.5 and 5.9 days respectively for four larval instars at 30 °C in wheat.

Table 1 : Duration of different life stages of *R. dominica*

Life Stages	**Life tenure (days)**	**Mean (days)**	**SE (±)**
Egg	5 - 9	7.09	0.54
Larval instars			
I instar	6 - 7	6.74	0.42
II instar	7 - 9	7.82	0.27
III instar	7 - 8	7.04	0.34
IV instar	8 - 10	9.27	0.52
Total larval duration	28 - 34	30.87	0.94
Pupa	7 - 8	6.73	0.26

The pupal period of 7 to 8 days (mean of 6.73 days) was recorded in the present study but the same was ranged from 3.7 to 5.4 days at 30 °C in wheat (Elek, 1994). The pupa is white in colour, having a retracted head and enlarged thorax. Abdomen has seven visible terga and appendages (external genitalia) terminally. The dorsal surface was covered by hairs and the tip was covered by the skin of pupa whose removal was inevitable for distinguishing the sexes.

Table 2 : Dimensions of head capsule, thorax and body length of different larval instars and pupae of *R . dominica*

Stages		Head capsule (mm)	Thorax width (mm)	Body length (mm)
Larval instars	I	0.114	0.175	0.625
	II	0.176	0.315	1.024
	III	0.275	0.650	2.058
	IV	0.425	0.826	2.900
Pupa		0.625	0.927	3.422

The pupae were distinguished by features as described by Potter (1935). Externally, the males and females resembled each other. The insect was dark reddish brown in colour, cylindrical and varied in size from 2.5 to 2.9 mm in length. The sex was determined based on the characteristic colouration on the ventral side, it was having uniform brown colour in females, pale yellowish colour in males, as reported by Stemly and Wilber (1966), the differentiation being clear in freshly emerged beetles.

The present investigation had revealed that the developmental period in finger millet was in the range of 40 to 51days with a mean of 44.69 days at 25 °C ± 1.02 °C and 70 ± 5.30 per cent relative humidity (Table 3).

Table 3 : Duration of different parameters of biology and sex ratio of *R. dominica*

Duration	Range	Mean	SE (±)
Total developmental period (Days) (Egg + Larval + Pupal)	40 – 51	44.69	1.26
Mating period (min)	51- 123	81.53	3.29
Pre ovipositional period (days)	6-9	07.40	0.25
Ovipositional period (days)	28-45	34.20	0.95
Number of eggs laid per female	20-44	28.57	0.87
Adult longevity without food (days)			
Female	6-11	09.65	0.14
Male	5-8	07.24	0.21
Adult longevity with food (days)			
Female	43-71	54.70	1.58
Male	37-54	43.20	0.98
Sex ratio (female: male)			
Laboratory culture	1 : 1.135		
Godown culture	1 : 1.542		

The total development period occupied, 40 to 51 days on finger millet, during January to April (24 to 28 °C and 68 to 75 % RH). These results are in close agreement with that of Howe (1950) who came out with 40 to 44 days in whole meal wheat flour at 28 °C and 70 per cent relative humidity.

A predominance of males was observed among population raised in the laboratory culture, female: male (1: 1.135) as well as in the collections from godown, female: male ratio of 1:1.542 was observed (Table 3). Similarly, Cline (1973) and Abdul-Matin (1974) reported a sex ratio of 55 (female): 59 (male) and 1: 1.32, respectively.

The present revelations on the positive phototropic nature of this pest, its strong attraction to light, and its activeness during noon hours, and readiness in taking to flight are in conformity with the observations of Potter (1935). The first mating took place 12 to 24 hours after the emergence of adults as observed by Thomson (1966). Copulation, as reported by Pajni and Gill (1974), normally took place at noon time when there was bright sunlight. Copulation period ranged from 51 to 123 minutes (average: 81.53 minutes). Repeated mating between a pair was not uncommon. The pre-ovipositional period (from emergence to first egg laying) was spread over 6 to 9 days with an average of 7.40 days which is slightly lesser than the period of 11.8 ± 1.3 days, reported by Abdul-Matin (1974). A period of intense feeding by the beetles, producing a lot of frass before egg laying was observed. The frass which has the same composition of the flour (Breese, 1960) is produced only as a response to fertilization.

The ovipositional period ranged from 28 to 45 days with a mean of 34.20 days. The beetles generally laid the eggs singly in the frass produced as a result of the feeding habit of the beetle on finger millet grains. The number of eggs laid varied considerably. Occasionally, the eggs were laid in groups of 2 to 5 as observed by Potter (1935) and Elek (1994).

Fecundity of 20 to 44 eggs (average of 28.57 eggs) per female, with the average egg production per female per day being 3.5 was observed.

Longevity of 6 to 11 days (mean of 9.65 days) in females and 5 to 8 days (mean of 7.24 days) in males was observed in the absence of food. The present investigation further revealed a longer adult life on finger millet grains (female: 43 to 71 days with a mean of 54.7 and male: 37 to 54 days with a mean of 43.2 days). Moreover, the female either in the presence or in absence of food lived longer than the males.

3.2. Infestation Behaviour and Nature of Damage of *R. Dominica*

Nature of damage of R. dominica

The newly emerged larva after hatching was very active and started crawling in search of suitable grain. After choosing the grain, it started scraping the grain or when the grain was a damaged one, it started boring the grain. Once it started feeding, the faecal pellets were used to bind into the adjacent grains tightly together. Like this, the larva formed a cluster of grains and was found feeding inside the cluster. Pupation also occurred inside the grain cluster. Later the adults emerged from the grain clusters. Newly emerged adults started scraping the grain coat and when white endosperm was seen, they started feeding on it. They started feeding voraciously leaving behind only seed coat.

Amount of food required for total larval development

The entire larval development occurred by feeding on broken finger millet grains. The maximum weight of finger millet required for larval development was 0.114 g and that of lowest was 0.071 g with an average of 0.096 g. Fig. 1 represent the variation in the amount of larval food requirement.

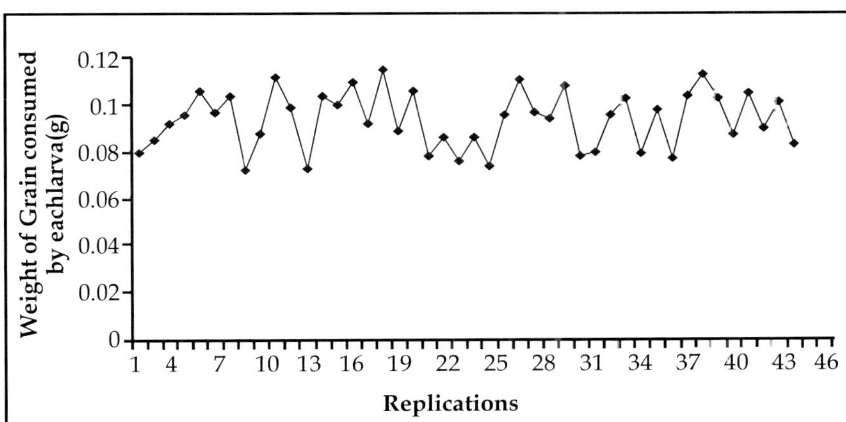

Fig 1 : Consumption pattern of finger millet by *R. dominica*
(Average consumption per larva = 0.096 g
45 larvae were maintained separately. One gram of finger millet was given to each larva)

The nature of larval development was quite different in finger millet from other cereals. Here in finger millet, larva started forming grain cluster and started feeding inside the grain cluster, pupation was also inside the grain cluster. Adult came out of the grain cluster through an irregular hole.

References

Abdul-Matin, A. S. M., 1974. Susceptibility of *Rhizopertha dominica* (F.) to ionizing radiation. *J. Stored Prod. Res.*, 10: 199-207.

Breese, M. H. 1960. The infestibility of stored paddy by *Sitophilus sasakii* (Tak.) and *Rhizopertha dominica (*Fab.). *Bull. Ent. Res.*, 51: 599-630.

Cline, L. D. 1973. Rearing of lesser grain borer on ground dog biscuits to obtain mixture larva and pupa. *Ann. Ent. Soc. Am.*, 66 (3): 551-552.

Elek, J. A.1994. Methods for collecting eggs and monitoring egg-hatch and immature development of *Rhizopertha dominica* (F.) (Coleoptera: Bostrichidae). *J. Stored Prod. Res.*, 30 (4): 261-265.

Howe, R.W. 1950. The development of *Rhizopertha dominica* Fab. under constant conditions. *Entomol. Mon. Mag.*, 86: 1-5.

Pajni, H. R. and Gill, K. M. 1974. Effect of light on the pests of stored products. *Bull. Grain Tech.*, 12 (2): 151-153.

Potter, C., 1935. The biology and distribution of *Rhizopertha dominica* (F.). *Trans. R. Ent. Soc. Lond.*,83: 449-479.

Stemley, P. G., and Wilbur, D. A., 1966. A colour characteristic for sexing live adult of lesser grain borer. *J. Econ. Entomol.*, 59 (3): 760-761

Thomson, V. 1966. The biology of the lesser grain borer, *Rhizopertha dominica* (Fab.) (Coleoptera: Bostrichidae). *Bull. Grain Tech.*, 4(4): 163-168.

— xxx —

9

The Need and Scope for Pest-free Grains in Storage

DAKSHINAMURTHY, A.
Director (Retd.), Indian Institute of Crop Processing Technology
Thanjavur-613 005, Tamil Nadu.

SUMMARY

The paper emphasizes the need and scope for insect free grains in storage while giving value of certain insecticides and bio pesticides. Final conclusion focuses role of the integrated pest management.

Key words: Grain Storage, Integrated Pest Management, Optimum Moisture

1. INTRODUCTION

The Green Revolution in India, has led to self sufficiency in food and also surplus food grains in storage. At national level, in the post harvest system, the loss in storage is quite high compared to other factors. In the field crops, the extent of infestation by insects may be higher, but the economic loss may be a fraction of the same. However, in storage of grains and seeds, the extent of infestation may be lesser, but the economic loss may be quite high. More than the quantitative loss, the substantial reduction in monetary value of the infested lot due to loss in quality is quite significant. The contamination of food grains due to immature stages of the insects, moulted skin, excreta, dead insects, reduction in nutrient content of grains, secondary infection by microbes, discolouration, unacceptable odour or off-smell are some of the problems associated with infested grains which make the food unacceptable and also unhygienic. Similarly, the loss in germination of seeds and reduction in vigour of the seedlings pose series of problems to the farmers, even threatening their subsistence and existence. The chemical control of insects in large scale storage of food grains using the insecticides like malathion, deltamethrin and fumigants like aluminium phosphide were quite convenient and effective. Due to the misuse of these insecticides and the inherent toxic effects, there is widespread awareness to minimize the insecticides in storage of grains. So there is an urgent need to find alternate means of insect pest management in storage of grains and of course there is scope to adopt an integrated system of pest management of practical utility.

2. GRAIN MOISTURE AND CLEANING FOR STORAGE

The moisture content of grains for harvest is about 20% which should be brought down to safe storage moisture by drying so as to enable it free from insect infestation. The optimum moisture content of different grains for safe storage vary to a great extent (Table 1). Apart from the moisture content, the initial quality of the produce to be stored is also very important. If the grain lot contains dirt, chaff and brokens, it becomes susceptible to insect attack. Hence, the use of grain cleaners before storage prevents the problem at a later stage.

Table 1 : Moisture levels for safe storage of food grains

Food Product	Moisture (%)
Paddy	14
All other cereals	12
Raw and Parboiled rice	14
Rice bran	10
Wheat flour	12.5
Rava / Sooji	13.5
Maida	12.5
Pulses	10-12
Oil seeds	6-8
Spices	10

3. FIELD INFESTATION

The insect in stored food grains is mostly due to the carry over of field infestation. The field infestation in rice was reported by Israel and Vedamoorthy (1956), in wheat by Simmons and Ellington (1927), in Sorghum by Carino and Morallo – Rajesus (1976) and in Cowpea by Hagstrum (1985). Preharvest spray of insecticides in rice and maize have helped to reduce the insect multiplication in storage from field carryover and consequent loss (Dakshinamurthy, 1991). Hence, a thorough estimation of infestation before storing the grains paves the way for pest free storage of grains at a later date. The procuring agencies have fixed certain norms for various quality parameters (Tables 2, 3 and 4) while purchasing the commodities. However, the

Table 2 : FAQ specifications for paddy

S. No.	Constituents	Maximum Limit (%)
1	Foreign matter	
	a) Inorganic	1.0
	b) Organic	1.0
2	Damaged, discoloured, sprouted and weevilled grains	3.0
3	Admixture of lower classes	10.0
4	Immature, shrunken and shrivelled grains	3.0
5	Moisture	18.0

Table 3 : FAQ specifications for rice

S. No.	Refractions	Maximum Limit (%) Grade 'A'
1	Brokens	
	Raw	25.0
	Parboiled	14.0
2	Foreign matter	0.5
3	Damaged / slightly damaged	2.0
4	Discoloured grains	3.0
5	Chalky grains	5.0
6	Red grains	3.0
7	Admixture of lower class	10.0
8	Dehusked grains (brown rice / under milled)	12.0
9	Moisture content	14.0

percentage of insect infestation by is based on damaged grains, which has not taken into account the hidden infestation the immature forms of insects. To assess the same, Ashman Simon hidden infestation detector, staining or irradiation methods may be followed.

Table 4 : Codex standards for rice

S. No.	Refractions	Maximum Limit (%)
1	Moisture content	15
2	Filth	0.1
3	Organic matter	0.5
4	Inorganic matter	0.1
5	Brokens	Buyers preference
6	Damaged kernels	3.0
7	Immature kernels	2.0
8	Chalky kernels	11.0
9	Red kernels	4.0

4. STORAGE STRUCTURES

Once, the grain is free from inoculum and of optimum moisture content, a suitable storage structure may be used to store for varying duration. The adoption of a metal storage structure or silo may be advantageous to prevent the quantitative and qualitative loss due to rodents. Though bulk storage of grains in metal or concrete silos is prevalent in many places, the most common practice to store the commodities is by use of gunny bags. As they are permeable in nature, the cross infestation from the infested old lot to the new lot is unavoidable.

5. STORAGE PEST MANAGEMENT

5.1. Insect Traps in Storage

The insect traps of various types are available to suit different capacities for domestic and bulk storage of grains. These traps were found effective for coleopteran insects rather than lepidopteran insects. The insect probe trap is ideal for domestic level small container storage and a similar trap of higher length can be used for bag storage. The pitfall trap can be used in bulk storage of grains while the insects are active on the surface of the grains.

5.2. Plant Products and Inert Dusts

At small scale level, mixing of plant products like *Acorus*, neem, pepper or *Vitex* in powder or other forms help to some extent as a deterrent (Dakshinamurthy, 1988). Even the use of inert dusts like ash or activated clay reduce the development of insects. In large scale storage, these products are not suitable as they are not economical or not preferred by the consumers.

5.3. Separation of Infested Grains

Periodical inspection to detect the infestation at an early stage, drying, sieving and blowing are the other means to protect the grains from further deterioration. The separation of infested wheat in bulk storage has been found to be useful to have quality wheat flour (Sahay and Dakshinamurthy, 1995).

5.4. Heat Treatment of Grains

Heat treatment of grains to a temperature of 60° C for a period of 15-20 min was found effective in disinfesting the grains. However, care should be taken to avoid the loss in quality of the grain products according to its end use.

5.5. Use of Carbon Dioxide

The role of carbon dioxide and its toxicity to storage insects was reported by Oosthuzien and Smidt (1942), Aliniazee (1971) and Banks (1975). At 20% carbon dioxide, the mortality was quite high in a period of 10 days. Similarly, an attempt was made to use biogas containing 40% carbon dioxide and 60% methane to control the storage insects of rice, wheat, maize and pulses and found effective (Palaniswamy and Dakshinamurthy,1986).

5.6. Fumigation and Newer Fumigants

As a curative measure, any infested product may be fumigated with proper fumigants like aluminium phosphide at the rate of 9 g/ tonne and exposed for 4-5 days. Recently some of the newer fumigants such as Ethylformate at the rate of 200 g/cubic metre with an exposure period of 48-72 h and sulphuryl fluoride 40 g/cubic metre for 24 h have been tried in large scale.

5.7. Use of Insecticides

The use of insecticides is in vogue in scientifically managed bulk grain storage godowns. However, the grains meant for consumption should be free from insects, insect fragments and insecticide residues.

Considerable amount of work has been done on the effect of direct mixing of insecticides with grains and their use in preventing the insect infestation in stored grains (Watters, 1977 and Muhihu, 1980). Even a grain treating machine to use malathion for uniform mixing with wheat was developed, but is not in use now. In fact, the direct mixing of insecticides with food grains is not allowed as per Prevention of Food Adulteration Act (PFA), 1954. However, for protection of seeds the mixing of insecticides can be continued and the seed treating equipment will be more useful.

Generally, the insecticides release vapours even though, the level may not be as high as that emanating from fumigants. Girish et al. (1973) reported that dichlorvos had a knock down effect within 50 – 60 minutes for stored grain pests. The lethal effect of dichlorvos was predominantly due to a mobile form and the vapour at toxic concentration had moved through grain up to 4 m using aeration (Desmarchelier *et al.,* 1977). The un-uniform application of dichlorvos and its effectiveness as vapour had also been used.

Impregnation of gunny bags with insecticides was found to reduce the insect infestation in storage (Morallo Rajesus *et al.,* 1975). The ability of synthetic pyrethroids such as deltamethrin as package treatment against stored product beetle has been recorded. Based on the efficacy of this method and to meet the large scale requirement of treated bags, a gunny bag impregnation unit was developed to mechanically treat the bags at a faster rate. However, use of the treated bags was discontinued due to the predicted health hazards for the workers engaged in handling, loading and unloading of the bags.

Thus, it is evident that integrated pest management is the appropriate solution for protection of grains from the insect infestation, loss in quantity and quality without filth in food commodities and insecticide residues.

References

Aliniazee, M.T. 1971. The effect of CO_2 gas alone or in combination. *J. Stored Prod. Res.,* 7. 243-252.

Banks, H.J. 1975. Recent advances in the use of modified atmosphere for stored product pest control. pp.198-217. Proc. II Int. work Conf. Stored Prod. Entomol. Ibadan, Nigeria,

Carino, F.O. and B. Morallo Rajesus 1976. Control of *Sitophilus zeamais* and *Rhizopertha dominica* by pre harvest sprays and stack treatment with three insecticides in three sorghum varieties. *Philipp. Agric.*, 60: 81-95.

Dakshinamurthy, A. 1988. Effect of certain plant products on storage pests of paddy. *Trop. Sci.*, 28: 119-122

Dakshinamurthy, A. 1991. Pre harvest spray of insecticides in maize for the management of *Sitotroga cerealella* in storage. *Bull. Grain Technol.* 29(1): 63-64

Desmarchelier, J.M., H.J. Banks, P. Williams and W.Minnett 1977. Toxicity of dichlorvos vapour to insects in aerated and non-aerated wheat and comparison of the vapour action of dichlorvos and malathion. *J. Stored Prod. Res.*, 13: 1-12.

Girish, G.K., R.K. Goyal and K.Krishnamurthy 1973. Some new organophosphorus insecticides against stored grain pests. *Bull. Grain Technol.*, 2: 59-67.

Hagstrum, D.W., 1985. Pre-harvest infestation of cowpeas by the cowpea weevil and population trends during storage in Florida. *J. Econ. Entomol.*, 78: 358-361.

Israel, P. and G. Vedamoorthy 1956. Origin of infestation of stored paddy. *Rice News Teller*, 4: 75-82.

Morallo-Rajesus, B., L.M.Varea and E.G.Nerova 1975. Insecticide impregnation of sacks and use of plastic lining for the protection of stored corn against insect damage. *Philipp. Agric.*, 59: 196-204.

Muhihu, S.K. (1980). Screening of new insecticides for use in farm level storage of maize in Kenya. *Keyna Entomologists Newsl.*, 4: 11.

Oosthuzien, M.T, and Smidt, U.W. 1942. The toxicity of CO_2 to the cowpea weevil. *J. Econ. Entomol. Soc.*, South Africa. 5-99.

Palaniswamy, P.T.,and A. Dakshinamurthy, 1986. Biogas to control rice storage pests. *Intl. Rice Res. Newsl.*,11(6): 25.

Sahay, K.M. and A.Dakshinamurthy, 1995. Separation technology for infested wheat in bulk storage. *J. Food.Sci.Technol.*, 32(3): 255-258

Simmons, P. and G.W.Ellington. 1927. Dispersion of Angoumois grain moth to wheat fields. *J. Agric. Res.*, 34: 459-467.

Watters, F.L. 1977. Comparison of acephate and malathion applied to stored grain for control of rusty grain beetles and red flour beetles. *J. Econ. Entomol.*, 70: 377-380.

— xxx —

10

Alternative Fumigants for Stored Grain Protection in India

RAJENDRAN, S.
Food Protectants and Infestation Control Department,
Central Food Technological Research Institute, Mysore-570 020

SUMMARY

In India, research has focused on chemical compounds such as allyl acetate, ethyl formate and sulfuryl fluoride and CO_2-rich atmosphere as alternatives. In 24 hr tests at e25^0 C, the most tolerant insect for allyl acetate and ethyl formate was *Tribolium castaneum* (150 g/m^3 and 140 g/m^3 for 100% mortality of all life stages, respectively) and *Sitophilus oryzae* L.for sulfuryl fluoride (40 g/m^3). The Khapra larvae, *Trogoderma granarium* were highly susceptible to sulfuryl fluoride (100% mortality achieved at the dose of 5 g/m^3). Following laboratory and field experiments with most tolerant stored grain insects, it was established that allyl acetate could be used for grain fumigation at farm level (< 1 tonne storage) at the dose of 200 g/m^3 for 72 hr, ethyl formate at 214 g/m^3 for 72 hr and sulfuryl fluoride at 40 g/m^3 with 24 hr exposure for bag-stacks of wheat at e25^0 C. Field trials with CO_2-rich atmosphere showed the possibility of its application for insect control and long term storage of bagged rice (Basmati and non-Basmati) with least effect on commodity quality.

Key words: Fumigation, Store Grain Pest, Tolerance Limit.

1. INTRODUCTION

India produces about 212 million tonnes of food grains annually. Losses in food grains occur at various stages from production till consumption by depredating agents, particularly insect pests. Fumigation plays a major role in insect pest elimination in food grains during storage. Currently, India uses about 470 tonnes of methyl bromide and 670 tonnes of aluminium phosphide tablets for pest control in various stored products including food grains. Methyl bromide is a broad-spectrum fumigant and the major advantage is its rapid action (exposure period 24-48 hr only) against the target pests. However, because of its ozone depleting potential, there has been a global move to phase out methyl bromide as per the Montreal Protocol 1992. In India, until July 2000, the fumigant was used for treating soil structures (e.g. flour mills) and stored commodities. Lately, the use methyl bromide has been banned except for quarantine and pre-shipment (QPS) treatments (MoEn, 2000).

Phosphine fumigant has several favourable qualities (eg. low cost, easy to apply, least residues and nil effect on treated commodities). However, development of resistance in stored grain insect pests to phosphine in the warehouses across the country is a serious problem for concern and control failures under field conditions been reported (Rajendran, 1999). Incidence of phosphine-resistant strains of stored grain insects has been reported in several other countries as well. Thus, the fumigants presently used are under pressure on environmental or biological grounds and hence, there is a need to look for alternative fumigants to ensure continued protection of valuable food commodities (Rajendran, 2001). Realizing the significance of the problem, intensive research on carbonyl sulfide, ethane dintrile, ethyl formate, methyl phosphine, ozone, sulfuryl fluoride and compounds from plant sources as alternatives have been carried out in different countries. In India, ethyl formate, allyl acetate, sulfuryl fluoride and CO_2-rich atmosphere have been studied for controlling stored grain insects. The present paper discusses about the salient features of these alternative fumigants and their possible role in stored grain protection.

2. ALTERNATIVE FUMIGANTS

2.1. Allyl Acetate

Allyl acetate, a colorless liquid with acrid odour is used in food and beverage industry as a flavouring agent at 1-5 ppm levels and is also used in the synthesis of various chemicals. Allyl acetate is insoluble in water, has low vapour pressure and is highly inflammable (Table 1). It is toxic to humans by ingestion, inhalation and contact; however, there are no reports indicating that the compound is carcinogenic.

Table 1 : Desirable properties of alternate fumigants

S. No	Properties	Allylacetate	Carbonyl sulfide	Ethylfor mate	Sulfuryl fluoride
1	Molecular weight	100.1	60.1	74	34
2	Boiling point (°C)	103.5	-50.2	54.1	-37
3	Vapour pressure (mm of Hg)	21	9,400	200	13,442
4	Flammability (% by volume)	1.1-7.5	12-28.5	2.8-16.5	Non flam mable
5	Water solubility (g/L)	Insoluble	1.4	14.5	0.75
6	TLV- TWA* (ppm)	100	10	100	5
7	LC50 for rats (ppm)	1000/hr	1070/4 hr	8000/4 hr	991/4 hr
8	Reaction with metals	Corrosive	Corrodes copper	Corrodes unpainted iron and steel	Nil
9	Effect on germination	Not known	May affect	No effect	No effect
10	Registration status	Nil	Australia (Expected)	Australia	Canada, USA, Europe

* Threshold limit value-Time weighed average concentration in air.

In bioassays against mix-age cultures of stored grain insects it was noted that the rust-red flour beetle, *Tribolium castaneum* was the most tolerant to allyl acetate whereas the lesser grain borer, *Rhyzopertha dominica* was highly susceptible that 100% mortality of all developmental stages of the insects was achieved at the dose of 150 g/m^3 and 50 g/m^3, respectively (Rajendran and Muralidharan, 2005). In further tests it was noted that the fumigant toxicity of allyl acetate could be enhanced by CO_2 at 10 and 20% levels. In small-scale fumigations, allyl acetate alone at the dose of 200 g/m^3 in 72 hr exposure proved effective in controlling stored grain insects in wheat (Table 2).

Table 2 : Most tolerant insects and commodity dosages for alternative fumigants

S. No	Fumigants	Tolerant species	Suggested dosage for wheat/raw rice
1	Allyl acetate	*Tribolium castaneum*	200 g/m^3, 72 hr
2	Ethyl formate	*Tribolium castaneum*	214 g/m^3, 72 hr
3	Sulfuryl fluoride	*Sitophilus oryzae*	40 g/m^3, 24 hr

Being a liquid fumigant, allyl acetate is particularly useful for disinfesting food grains and other commodities stored in small quantities (<1 ton). Phosphine, due to its high vapor pressure (31,930 mm of Hg at 30^{0}C) leaks out through the storage structures traditionally used by the farmers. In this context, allyl acetate having low vapor pressure (21 mm of Hg) may be suitable for use at farm level. A formulation containing allyl acetate + CO_2 could be useful for treating grain in larger quantities (Leelaja *et al.,* 2007).

2.2. Ethyl Formate

It is a low boiling liquid fumigant with a pleasant odour. It is naturally present in certain raw and processed foods at low levels (0.05 to 1 mg/kg). Ethyl formate has been reported to show rapid action against insect pests and is known to break down into natural products, ethanol and formic acid. Ethyl formate is generally recognized as a safe fumigant It has been used for individual package fumigation of dry fruits in Australia, South Africa and other countries. In Australia, Vapormate®, a formulation containing 16.7% ethyl formate in liquid CO_2 has been developed for insect control in food processing facilities (dosage 420 g/m^3, 6 hr exposure) and in stored grain (420 g/m^3, 24 hr exposure) (Ryan, 2005). CO_2 serves the dual function of reduction in flammability risk of ethyl formate and synergistic effect.

Extensive laboratory and field experiments on ethyl formate conducted by the Central Food Technological Research Institute (CFTRI) revealed effective doses ranging 300-400 g/m^3 with a 72 h exposure period for treating different cereals, pulses, spices, dry fruits and tree nuts (Muthu *et al.,* 1984). However, recent experiments by the author on 160 tonne wheat stacks at e"27^0 C showed that a dose of 214 g/m^3 for 72 hr was adequate.

2.3. Sulfuryl Fluoride

Sulfuryl fluoride is a colourless, odorless gas available in steel cylinders as a liquid (99.8%) under pressure. It doesn't cause cancer and other health effects. It is environmentally safe also. The fumigant can be used under normal atmospheric pressure as well as under vacuum. It is non-corrosive to electronic equipments and does not taint the commodities. Sulfuryl fluoride under the trade name 'Vikane' has been used for fumigating structures and furniture to control drywood termites and wood destroying pests in USA, since 45 years. The EPA, USA has recently approved the use of sulfuryl fluoride (trade name 'ProFume') for insect pest control in stored cereals, dried fruits and tree nuts and food storage facilities. ProFume has been approved for use in food commodities and food processing facilities in Canada and in the Europe (Prabhakaran, 2006). The EPA has fixed tolerance limits in food commodities for sulfuryl fluoride residues (e.g. wheat - 40 ppm as fluoride and 0.1 ppm as sulfuryl fluoride). Apart from USA, China has been producing sulfuryl fluoride (trade name "Xunmiejin") since 1983 (Guogan *et al.*, 1999).

Sulfuryl fluoride has vapour pressure nearly seven times higher (13,442 mm of Hg at 25°C) than that of methyl bromide (1824 mm of Hg). Therefore it is more penetrative into treated commodities than methyl bromide. Egg stage of all insects is tolerant to sulfuryl fluoride. Bell *et al.* (2006) reported that egg tolerance could be overcome by increasing the exposure period and by raising the treatment temperature. It is expected that sulfuryl fluoride will play a significant role in quarantine treatment of food commodities in the near future.

In India, at CFTRI, for the first time, studies were conducted on sulfuryl fluoride. Toxicity tests show that Khapra larvae *Trogoderma granarium* were quite susceptible to the fumigant, whereas, eggs of *Oryzaephilus surinamensisI* L. and *Sitophilus oryzae* L.were highly tolerant. A dosage of 40 g/m^3, with 24 hr exposure period was necessary to achieve 100% mortality of all life stages of the most tolerant insects in grain stack treatments (Table 2).

2.4. Controlled Atmospheres

CO_2 is toxic to stored grain insects at e''35% levels in air. Krishnamurthy *et al.* (1993) developed simple methods for on-site generation of CO_2, based on the reaction of naturally available limestone and dilute hydrochloric acid for application in airtight storages of <1 tonne capacity. Large-scale CO_2 fumigation trials were conducted at the Food Corporation of India Buffer storage complex, Mumbai on milled raw rice stacks (126 and 147 tonnes each) and in a concrete silo filled with 2200 tonnes of wheat (Narasimhan *et al.*, 1993). Complete mortality of all stages of resident infestation of *Tribolium castaneum* Hbst., *Oryzaephilus surinamensis, Ephestia cautella* Walker and psocids was achieved in rice stacks following CO_2 treatment. There was no change in colour or odour of rice and wheat. Rajendran *et al.* (2002) demonstrated storage of bagged Basmati rice (30 tonne stacks), under CO_2-rich atmosphere for 3 months. Insects were completely controlled and there was no adverse effect on the quality of Basmati rice.

Disinfestation of food grains by CO_2 treatment requires a long period of 10 days and above. However, this drawback has been overcome in developed countries by raising the treatment temperature to 30-40^0 C or by applying CO_2 under high pressure of 20-25 bars.

2.5. Others

2.5.1. Carbonyl sulfide

The fumigant properties of carbonyl sulphide, an industrial gas, have been elaborately studied in Australia and it has been patented (MoEn, 1993). Carbonyl sulfide is moderately toxic when compared with methyl bromide or phosphine. It is claimed that it can achieve satisfactory insect control in a period of 6 hr to 7 days. Studies in Australia show that the fumigant does not affect the quality of treated wheat and germination. However, investigations in China indicate that carbonyl sulfide could cause off-odour problem in treated commodities and affects germination also. It is suspected that hydrogen sulfide present as an impurity in carbonyl sulfide is responsible for tainting in treated commodities. Carbonyl sulfide is likely to be registered as a grain fumigant in Australia under the trade name 'Cosmic'.

2.5.2. Ethane dinitrile

Ethane dinitrile or cyanogen (boiling point -21.2^o C) is a colourless gas with an almond-like odour. Its TLV is 10 ppm. CSIRO Entomology, Australia, has patented the fumigant. It has got potential application for the control of wood borers, termites and stored grain insects. Ethane dinitrile is highly phytotoxic and hence it will be useful for weed control and revitalization of grains. In laboratory tests at CFTRI it was noted that Khapra larvae and the developmental stages of *Sitophilus oryzae* were highly tolerant that they require doses of e"32.8 g m^{-3} for 48 hr to achieve 100% mortality. The fumigant is likely to be registered in Australia under the trade name 'Sterigas'.

2.5.3. *Plant products*

World over, interest has been shown on plant products (essential oils and their components) for fumigant action since it is believed that compounds from plant sources may have the advantage over conventional fumigants in terms of low mammalian toxicity, rapid degradation and local availability. More than 75 native plant species belonging to different families have been tested at laboratory level for fumigant toxicity. The active components showing insect toxicity in the vapour phase can be grouped as monoterpenoids, cyanohydrins and cyanates, sulphur compounds (dimethyl disulphide, diethyl trisulphide, di-n-propyl disulphide, allyl disulphide, diallyl trisulphide, allyl thiosulfinates), alkaloids (Z-asarone) and others (methyl salicylate, benzene derivatives, bornyl acetate, terpinolene). Most of the active ingredients are the secondary metabolites secreted in plants as chemical defense against pest organisms. Essential oils or their components are generally less toxic to insects when compared with conventional fumigants. Fumigants from plant sources lack the most important property of an ideal fumigant i.e., sufficient vapour pressure for diffusion and penetration in to commodities to kill the pests. Plant fumigants may or may not affect seed germination.

Apparently, compounds of plant origin can be used only for small-scale applications or for space treatments. They need some carriers (eg. CO_2) for even distribution and penetration into the commodities or they can act as adjuvants for conventional fumigants. In India, Shukla *et al.* (2004) reported that an herbal fumigant formulation using *Eucalyptus camaldulensis* Hook. Oil was suitable for stored grain insect control.

3. CONCLUSION

Although alternatives are available for effective grain fumigation, they do have some limitations. Allyl acetate has low vapour pressure that it can be used only for small-scale fumigations. Ethyl formate preferably needs a carrier gas for effective distribution inside grain mass. Sulfuryl fluoride is not manufactured in India and insect eggs are exceptionally tolerant to the fumigant. Use of CO_2-rich atmosphere is economical only for long-term storage of food grains but not for routine disinfestations.

Acknowledgement

Some of the data presented in the paper are from studies carried out under the project "Substitutes for methyl bromide in pre-shipment fumigation of food commodities" sponsored by the Department of Agriculture and Cooperation, Ministry of Agriculture, Government of India, New Delhi,for whom I owe my sincere thanks.

References

MoEn, 1993. Australian storage scientists patent new fumigant. ACIAR Post harvest Newsletter 27, December 1993, p.2

MoEn,2000. Ozone depleting substances (Regulation and Control) Rules, 2000. The Gazette of India. Extraordinary Part–II. Section 3 sub–section (ii) No. 471. Ministry of Environment and Forests, Notification dated 19th July 2000.

Bell, C.H., 2006. Factors affecting the efficacy of sulfuryl fluoride as a fumigant. pp.519-526 In: Lorini, I. et al (Eds.), Proceedings of the 9th International Working Conference on Stored Product Protection, 15-18 October 2006, Campinas, Brazil, Brazilian Post-harvest Association, Campinas, Brazil.

Guogan, X., Zhongmei, C., Zhao, S., Nengzhi, Q., 1999. The development of sulfuryl fluoride (SO2F2) in China- a brief introduction pp. 562-566. In: Jin Zuxun, Liang Quan, Liang Yongsheng, Tan Xianchang and Guan Lianghua, (Eds.), Stored product Protection, Proceedings of the 7th International Working Conference on Stored Product Protection, 14-19 October 1998, Beijing, China, Sichuan Publishing House of Science & Technology, Chengdu, China.

Krishnamurthy, T.S., Muralidharan, N., Krishnakumari, M.K., 1993. Disinfesting food commodities in small storages using carbon dioxide rich atmospheres. *Intl. Pest Control.*, 35, 153-156.

Leelaja B.C., Rajashekhar Y., Vanitha Reddy P., Khamrunissa Begum., Rajendran S., 2007. Enhanced fumigant toxicity of allyl acetate to stored - product beetle pests in the presence of carbon dioxide. *J. Stored Prod. Res.*, 43, 45-48.

Muthu, M., Rajendran, S., Krishnamurthy,T.S., Narasimhan,K.S., Rangaswamy, J.R., Jayaram, M., Majumder, S.K., 1984. Ethyl formate as a safe general fumigant. pp. 369 393. In: Ripp, B.E. (Ed.), Controlled Atmosphere and Fumigation in Grain Storages. Proceedings of an International Symposium on Practical Aspects of Controlled Atmosphere and Fumigation in Grain Storages, 11-22 April 1983, Perth, Western Australia. Elsevier, Amsterdam.

Narasimhan, K.S., Rajendran, S., Krishnamurthy, T.S., Chandy, Z., Khedkar, P.M., Hakeem, M.A., 1993. Carbon dioxide fumigation trials in India pp. 37-49. In Navarro, S. and Donahaye, E. (Eds.),: Proceedings of an International Conference on Controlled Atmosphere and Fumigation in Grain Storage, 11-13 June1992, Winnipeg, Canada, Agricultural Research Organization, Bet Degan, Israel.

Prabhakaran, S., 2006. Commercial performance and global development status of Profume® gas fumigant. In: Proceedings of the 9th International Working Conference on Stored Product Protection, Lorini, I. et al. (Eds.), 15-18 October, 2006, Campinas, Brazil, Brazilian Post-harvest Association, Campinas, Brazil. pp.635-641.

Rajendran, S., 1999. Phosphine resistance in stored grain insect pests in India pp. 635-641. In: Jin Zuxun, Liang Quan, Liang Yongsheng, Tan Xianchang and Guan Lianghua, (Eds.), Stored Product Protection: Proceedings of the 7th International Working Conference on Stored-product Protection, 14-19 October 1998, Beijing, China, Sichuan Publishing House of Science & Technology, Chengdu, China.Rajendran, S., 2001. Alternatives to methyl bromide as fumigants for stored food commodities. *Pesticide Outlook*, 12, 249-253.

Rajendran S., Bhashyam, M.K., Muralidharan, N., 2002. Storage of basmati rice under carbon dioxide-rich atmosphere. *J.Food Sci. Tech.,* 39, 639-643.

Rajendran, S., and N. Muralidharan, 2005. Effectiveness of allyl acetate as a fumigant against five stored grain beetle pests. *Pest Management Science,* 61, 97-101.

Rajendran, S.,and Sriranjini, V., 2007 Plant products as fumigants for stored propduct insect control. *J. Stored Prod. Res.* (Accepted for pub lications)

Ryan, R., 2005. Ethyl formate. *Fumigants and Pheromones Newsletter,* 74: 5.

Shukla, A.C., Shukla, A., Mishra, R.C., Dikshit, A., 2004. Broad- spectrum herbal fumigant formulation for the management of stored grain pests. Paper presented in the International Conference on Controlled Atmosphere and Fumigation in Stored Products, 8-13 August 2004, Gold Coast, Australia.

— xxx —

11

Role of Plant Products and Oils in the Management of Pulse Beetle, *Callosobruchus chinensis* Linn.

RAMAZEAME, L.[1] AND D. ADIROUBANE[2]
[1] PG Student [2] Professor
Pandit Jawaharlal Nehru College of Agriculture and Research Institute
Karaikal - 609 603, U.T. of Puducherry

SUMMARY

Coconut oil (*Cocos nucifera* L.), Pongam oil (*Pongamia glabra* L.), Mahua oil (*Madhuca indica* J.), Castor oil (*Ricinus communis* L.), Sweet flag rhizome powder (*Acorus calamus* L.), Neem kernel powder (*Azadirachta indica* An.dr.Juss.), Notchi leaf powder (*Vitex negundo* L.) and Adathoda leaf powder (*Adathoda vasica* Medi.) were tested for their efficacy against pulse beetle in comparison with the activated clay for long term storage *i.e*, upto 10 months. All the treatments had suppressed the beetle damage. Neem kernel powder was exerted highest percent of reduction while Adathoda leaf powder had lowest percent reduction over control.

Key words: Plant Products, Management, Pulse Beetle

1. INTRODUCTION

The pulse beetle *Callosobruchus chinensis* L. is reported to be the major pest infesting all types of pulses both in the field and in storage. The differential rate of damage by *C. chinensis* in different pulses was reported to be 68 per cent in cowpea, 56 per cent in Bengal gram, 49 per cent in red gram and 52 per cent in green gram over a storage period of 6 months (Sharma, 1984). The grain damage was as high as 69.93% under storage condition (Singh *et al.,* 2001). This problem is serous at small farmers' level, village traders and average households where storage conditions are poor and inadequate. To control the pulse beetle in storage, a number of synthetic organic insecticides such as malathion and dicholorvos have been recommended. The admixture synthetic insecticide with food grains has more recently been banned in many countries. There were also reports that the pulse beetle developing resistance to malathion (Singh and Srivastava 1983). Plant products traditionally used against pulse beetle appear to be quite safe and promising. Several authors have reported the insecticidal action and growth inhibiting effects of plant products on *C. maculatus* (Pandey and Singh, 1995). Botanicals and vegetable oils can be used to keep the stored pulse free from pulse beetle attack. Therefore, studies were undertaken to develop a suitable management technique with plant products for the pulse beetle.

2. MATERIALS AND METHODS

Studies on role of plant products and oils in the management of pulse Beetle, *C. chinensis* were conducted in the Entomology laboratory of the Pandit Jawaharlal Nehru College of Agriculture and Research Institute, Karaikal, U.T. of Pondicherry during 2004-2005.

2.1 Mass Culturing of Pulse Beetle

Pulse beetles were mass reared on green gram (local variety) in the laboratory. The moisture content of the grains was adjusted to 11.0 per cent by sun drying. The mass culturing was initiated by confining 10-20 freshly emerged beetles in the plastic containers of 59x21x18 cm having 500 g of green gram and it was covered with kada cloth and secured tightly with rubber band. Such containers were stacked in iron shelves. Mass culturing of *C. chinensis* was done at room temperature in the plastic container and adults were collected for the study.

2.2. Identification of the Plant Products and Oils

Weeds, medicinal plants, edible and non edible oils which have insecticidal and ovicidal values were identified for this study. The plant materials collected, were shade dried and made into powder in a grinder. The concentration of botanical powder in all the cases was uniform and maintained at 5 g / Kg. The vegetable oils and activated clay were collected from the local market. The concentration of vegetable oils in all the cases was uniform and maintained at 5 ml/ Kg. But in the case of activated clay the dosage was 10 g / Kg. The seeds of black gram ADT 3 were sun dried to the moisture content of 9 per cent. Healthy seeds were separated and selected for the experimental purpose. The details of treatments along with dosage are furnished in Table1.

Table 1 : Details of the Treatments

Treatments	Vegetable products used	Scientific name of the plants	Dose/kg tested
T1	Coconut oil	*Cocos nucifera* (L.)	5ml
T2	Pongam oil	*Pongamia glabra* (L.)	5ml
T3	Illuppai oil	*Madhuca indica* (J.)	5ml
T4	Castor oil	*Ricinus communis* (L.)	5ml
T5	Vasambu rhizome powder	*Acorus calamus* (L.)	5g
T6	Neem kernel powder	*Azadirachta indica* (An.dr.Juss)	5g
T7	Notchi leaf powder	*Vitex negundo* (L.)	5g
T8	Adathoda leaf powder	*Adathoda vasica* (Medi.)	5g
T9	Activated clay	-	10g
T10	Untreated Control	-	-

2.3. Efficacy of Plant Products and Oils

From each treatment 100g of seeds were drawn from botanicals and oils treated healthy ADT 3 black gram seeds and were placed in plastic containers of 59 x 21 x 18 cm size, which were covered with plastic cover lid and used. Each of 100 g seeds was placed into small size plastic containers and covered with kada cloth and secured with rubber band and kept them stacked in iron shelves. Each treatment was replicated thrice. Ten pairs of newly emerged adults were released in each replication. The same procedure was adopted for all the treatments. After the 30th day the percentage of bore holes was worked out. The experiment was continued upto 10 months with an interval between each study was 30 days.

3. RESULTS AND DISCUSSION

3.1. Efficacy of Plant Products and Oils Against Pulse Beetle, *C. chinensis*

The efficacy of the plant products and oils was expressed by the number of seeds present with bore holes (Table 2). Since the plant products and oils were directly mixed with the seeds the adult beetle came in to contact directly with the plant products and oils. Among all the plant products, Adathoda leaf powder had maximum bore holes of 32.66 per cent, whereas neem kernel powder was highly effective during the entire study period *i.e.*, from May 2005 to February 2006.However the bore holes present in the pulses seeds gradually increased every month. The neem kernel powder was the best among treatments. There was an increasing trend in per cent bore hole invariably in all the treatments (Table 2).

Among the botanicals and oils, neem kernel powder had with highest reduction over control (99.51%) while adathoda leaf powder was with lowest reduction over control (98.15%). The percentage of reduction of bore holes over the control one month after treatment ranged between 99.51and 98.15. An increasing trend of infestation was recorded due to reduction in efficacy of plant products and oils and that could be observed over a period of ten months. However, the treatments were superior to control, which could be evinced from the population reduction of beetle in treated pulse seeds.

Though the results obtained have shown that the pulse grain could be protected from *C. chinensis* by treating the seeds with plant products and oils but their efficacy was slowly reduced and it was indicated by the increase in the per cent bore hole. This observation was in consonance with the findings of Tripathy *et al.* (2001) Singh and Yadav (2003) Tandon *et al.* (2004) and Allotey (2004). They found that the active principle present in the botanicals incited ovicidal activity, repellent principle protected the pulse seeds treated with different plant products and oils. This could be evidenced from the findings of Tripathy *et al.* (2001), Singh (2003), Ramamurthy *et al.* (2002) and

Table 2 : Efficacy of various plant products against *Callosobruchus chinensis* L.

S. No.	Treatments	Per cent bore hole (Mean of three replications) at various months										Over all mean
		1	2	3	4	5	6	7	8	9	10	
T1	Coconut Oil (*Cocos nucifera*)	1.00 (5.72) a	2.29 (8.70) de	10.28 (18.70) e	10.18 (18.61) ef	15.85 (23.46) de	20.69 (27.05) g	22.36 (28.22) ef	21.93 (27.92) f	24.84 (29.90) e	30.05 (33.24) de	15.94
T2	Pongam Oil (*Pongamia glabra*)	0.58 (4.36) a	1.28 (6.47) abc	4.97 (12.88) bc	6.77 (15.08)bc	9.93 (18.37) abc	12.62 (20.81) c	15.37 (23.08) b	17.10 (24.43) b	18.32 (25.34) bc	24.74 (29.83) b	11.16
T3	Mahua oil (*Madhuca indica*)	0.73 (4.90) a	1.80 (7.70) bcd	7.04 (15.38) d	8.25 (16.69) cd	13.56 (21.60) b-e	16.16 (23.70) e	18.41 (25.40) d	19.30 (26.06) d	21.30 (27.48) d	27.37 (31.54) c	13.39
T4	Castor oil (*Ricinus communis*)	0.89 (5.41) a	1.96 (8.04) cde	9.20 (17.65) e	9.24 (17.69) de	15.01 (22.79) cde	18.93 (25.79) f	21.32 (27.49) e	20.90 (27.20) e	22.70 (28.45) d	28.26 (32.11) cd	14.84
T5	Sweet flag powder (*Acorus calamus*)	0.62 (4.53) a	1.40 (6.78) a-d	6.10 (14.30) cd	7.34 (15.72) bc	11.24 (19.58) ad	15.13 (22.89) d	17.04 (24.38) c	18.26 (25.30) c	19.25 (26.02) c	26.86 (31.22) c	12.32
T6	Neem Kernel Powder (*Azadirachta indica*)	0.45 (3.83) a	0.85 (5.29) a	3.29 (10.45) a	5.13 (13.09) a	7.61 (16.01) a	9.92 (18.35) a	11.45 (19.78) a	14.34 (22.25) a	16.17 (23.71) a	17.14 (24.46) a	8.63
T7	Notchi leaf powder (*Vitex negundo*)	1.05 (5.87) a	2.96 (9.89) ef	12.30 (20.53) f	11.33 (19.67) f	16.69 (24.11) de	21.22 (27.43) g	23.29 (28.85) f	22.44 (28.27) f	27.34 (31.52) f	31.26 (33.99) ef	16.98
T8	Adathoda leaf powder (*Adathoda vasica*)	1.70 (7.40) a	3.62 (10.97) f	14.34 (22.25) g	15.17 (22.92) g	18.45 (25.44) e	22.56 (28.35)	25.58 (30.38) g	26.76 (31.15) g	27.65 (31.72) f	32.66 (34.85) f	18.84
T9	Activated clay	0.50 (4.07) a	1.05 (5.87) ab	4.24 (11.88) ab	5.94 (14.11) ab	8.37 (17.18) ab	11.09 (19.45) b	14.70 (22.55) b	15.16 (22.91) a	17.12 (24.44) ab	18.69 (25.61) a	9.68
T10	Control	92.30 (74.79) b	80.90 (64.18) g	81.72 (64.75) h	95.73 (78.36) h	84.80 (67.81) f	96.16 (78.71) i	95.54 (77.89) h	95.57 (77.88) h	97.29 (80.73) g	98.44 (83.06) g	91.84
	Overall Mean	9.98	9.81	15.34	17.50	20.15	24.44	26.50	27.17	29.19	33.54	-
	CD (p=0.05)	(3.85)**	(1.88)**	(1.47)**	(1.79)**	(4.39)**	(0.49)**	(0.99)**	(0.70)**	(1.33)**	(1.25)**	-

Treatment CD value at 0.05 per cent level: 0.65
Months CD value at 0.05 per cent level 0.65.
Treatments and Months interaction CD value at 0.05 per cent level 2.07

Values in the parantheses are arc sine transformed values
Mean in a column followed by same letter are significantly not different
**Significant at 0.05 per cent level

Allotey (2004), who reported that the efficacy lasted for a short period only. The reason for the reduction in the population of the pulse beetle could be due to oviposition deterrent activity of the different oils and plant products.

References

Allotey, J. 2004. Some aspects of the biology and control of *Callosobruchus maculatus* (F.) on some stored soybean *Glycine max* (L.) Merr varieties. *African J. Food, Agric. Nutrition and Development*. 4, (2).

Pandey, N.K. and S.C. Singh. 1995. Effect of neem leaf powder on survival and mortality of pulse beetle, *Callosobruchus chinensis* (L.) infestation in gram. *Uttar Pradesh J. Zool.,* 3: 162 – 164.

Ramamurthy, R., V. Rajaram and Land M.Rajendran. 2002. Efficacy of plant products against the storage pest bruchids in black gram. *Legume Res.*, 32. (342).

Sharma, 1984. Review of literature on the losses caused by *Callosobruchus* species (Bruchidae; Coleoptera) during storage of pulses. *Bull. Grain Tech.* 2(1): 62-71.

Singh, R., B. Singh and R.A. Verma. 2001. Efficacy of different indigenous plant products as grain protectants against *Callosobruchus chinensis* (Linn) on pea. *Indian J. Ent.*, 63(2): 179-181.

Singh, R.P and B.G. Srivastava. 1983. Alcohol extracts of neem (*Azadirachta indica*) seed oil as oviposition deterrent for *Dacus cucurbitae. Indian J. Ent.,* 45: 497-498.

Singh, S.C. 2003. Effect of neem leaf powder in infestation of the pulse beetle *Callosobruchus chinensis* in stored khesari. *Indian J. Ent.*, 65(2): 188-192.

Singh, V. and D.S. Yadav. 2003. Efficacy of different oils against pulse beetle, *Callosobruchus chinensis* in green gram *Vigna radiata* and their effect on germination. *Indian J. Ent.*, 65(2):281-286.

Tandon, S., A.K. Pant and Balairam. 2004. Effect of some plant extracts as grain protectants against pulse beetle *Callosobruchus chinensis*. L. *Seed Res.*, 32 (2); 197-199.

Tripathy, M.K., P. Sahoo, B.C. Das and S.Mohanty. 2001. Efficacy of botanical oils, plant powders and extracts against *Callosobruchus chinensis* Linn. attacking black gram (CV. T9). *Legume Res.*, 24(2): 82-86.

— xxx —

12

Effect of *Piper nigrum* L. Seed Powder and its Dust Formulations on pulse beetle, *Callosobruchus maculatus* (F.) (Coleoptera: Bruchidae)

GOVINDAN. K,[1] S. JEYARAJAN NELSON [2] AND P.M.M.DAVID [2]

[1] Ph.D., Scholar, Department of Agricultural Entomology, Tamil Nadu Agricultural University, Coimbatore, 641 003

[2] Professors, Department of Agricultural Entomology, Agricultural College and Research Institute, Killikulam- 628 252, Tamil Nadu

SUMMARY

Results on the effect of *Piper nigrum* L. seed powder and its dust formulations against *Callosobruchus maculatus* F. showed that *P. nigrum* seed powder 0.25 per cent registered 100 per cent mortality.. 15 hr after the treatment *P. nigrum* 10D (talc as filler) (Pn10DT) 4.00 per cent caused 100 per cent mortality at 84 hr after the treatment. *P. nigrum* 10D (fly ash as filler) (Pn10DF) 4.00 per cent caused cent per cent mortality at 72 hr after the treatment. *P. nigrum* 2.00 per cent seed powder acted as the best ovipositional deterrent. Adult emergence ranged between 6.66 (*P. nigrum* 1.50 per cent) and 16.00 (*P. nigrum* 0.25%). Seed weight loss was less (0.18%) in 10D 4.00 per cent. Pn10DF 4.00 per cent was most effective in reducing the oviposition. Egg hatchability of *C. maculatus* ranged between 57.12 (Pn10DF 2.00 %) and 73.73% (Pn10DF 1.00 %).

Key words: Seed Powder, Dust Formulations, *Callosobruchus maculatus*.

1. INTRODUCTION

Pulses are an important part of the vegetarian diet of Indian sub continent, being a rich source of protein (20–30 per cent). Major constraints in the production of pulses are due to the pest infestations in the field and in storage. Infestation arising in the field causes serious damage to the seeds during storage (Hagstrum, 1985; Dongre *et al.*, 1993). The bruchid, *Callosobruchus maculatus* is known to inflict quantitative and qualitative losses to stored pulses (Adams and Schulten, 1978). Caswell (1981) reported a loss of approximately 50% in cowpea in storage for 3 to 4 months due to infestation by *C. maculatus*.

Use of plant products as grain protectants is an age-old practice (Weaver *et al.*, 1992) appears promising (Jilani *et al.*, 1998). Various plant powders and their extracts have been reported to possess insecticidal and repelle nt activity against the bruchids (Agarwal *et al.*, 1988; Babu *et al.*, 1999). In the present study, *P. nigrum* seed powder and its dust formulations were evaluated for action on *C. maculatus*.

2. MATERIALS AND METHODS

Laboratory experiments were carried out at Agricultural College and Research Institute, Killikulam during 2003 - 2005. *C. maculatus* was reared on black gram seeds in glass jars covered with muslin cloth by following the method developed by Credland and Wright (1989). *P. nigrum* seeds were procured, shade dried, powdered using a pulverizer and used. Twenty grams of black gram seeds at 12 per cent moisture content were taken in petridishes. The *P. nigrum* seed powder and 10 % dust formulation (w/w) were added to black gram seeds and shaken thoroughly. Thirty newly emerged adults were released into each petridish and kept in the laboratory. Mortality was recorded at 24 hr interval for seven days. Three replications were maintained in each treatment. *P. nigrum* seed powders were made as 10 per cent dust formulations using different fillers *viz.*, talc and fly ash (waste product from thermal power station). The effects of the new dust formulations *viz.*, *P. nigrum* 10 D (talc as filler) (Ac 10DT) and *P. nigrum*

10 D (fly ash as filler) (Ac 10DF) were tested against *C. maculatus* at different concentrations *viz.*, 0.25, 0.50, 0.75, 1.00, 1.25, 1.50, 1.75 and 2.00 per cent as per the above procedure.

Oviposition, adult emergence, seed damage and seed germination. Twenty grams of black gram seeds were taken in glass bottles. *P. nigrum* seed powder and 10 % dust formulation (w/w) were added to the black gram seeds and shaken thoroughly. Activated clay (1:100) was used as a check. Then the glass bottles were covered firmly using muslin cloth. Five pairs of newly emerged adults of *C. maculatus* were released in to each glass bottle, covered firmly and kept in laboratory at ambient conditions. Three replications were maintained for each treatment. Number of eggs laid on the seeds was counted on third day after the release of the beetles. On 5th day after the release counts on translucent unhatched eggs and opaque hatched eggs were taken and hatchability percentage was worked out. On 15th day after the release of beetles all the dead insects were removed from the bottles to prevent them from getting mixed with the first generation (F1) off springs. The number of newly emerged adults was counted and removed from the bottle once in three days till the complete emergence of F1 offspring (up to 60 days after treatment). Observations on number of damaged and undamaged seeds were taken on 45th day after the treatment and percent seed damage was worked out. The weight of the seed was taken on 45th day after the treatment and seed weight loss percentage was worked out. After the experiment germination of black gram seeds was assessed on 90th day after the treatment using roll towel method.

3. RESULTS AND DISCUSSION

Per cent adult mortalities of *C. maculatus* after treatment were shown in Table-1 at hourly intervals upto 15 hrs. The highest mortality was observed in *P. nigrum* 2.00 per cent (41.11%) followed by *P. nigrum* 1.75 per cent (36.66%) and these two treatment were on par in their effect at 1 hr after treatment whereas no mortality was observed in *P. nigrum* 0.25, 0.50, 0.75 and 1.00 per cent. The same trend was observed up to 9 hr after the treatment. All the beetles were dead in *P. nigrum* 1.75 and 2.00 per cent at 10 hr after the treatment while in *P. nigrum* 1.00, 1.25 and 1.50 per cent 94.44, 93.33 and 92.22 per cent beetles were dead respectively. Similar findings of Aslam *et al.* (2002) who reported that black pepper, *P. nigrum* caused 100 per cent mortality of *C. chinensis* after 3.75 days of the treatment. At 12 hr after the treatment cent per cent mortality was seen in *P. nigrum* 1.00, 1.25 and

Table 1 : Insecticidal action of *Piper nigrum* seed powder on *Callosobruchus maculatus*

S. No.	Treatments	Per cent bore hole (Mean of three replications) at various months															Mean
		1 hr.	2 hr.	3 hr.	4 hr.	5 hr.	6 hr.	7 hr.	8 hr.	9 hr.	10 hr.	11 hr.	12 hr.	13 hr.	14 hr.	15 hr.	
1	*P. nigrum* 0.25 %	0.00 (0.52)	0.00 (0.52)	3.33 (10.51)	15.55 (23.03)	23.33 (28.84)	40.00 (39.22)	51.10 (45.63)	52.21 (46.27)	61.11 (51.42)	58.257 (52.25)	85.55 (68.12)	91.10 (73.20)	94.44 (78.69)	97.77 (82.80)	100.00 (89.47)	52.51 (46.43)
2	*P. nigru* 0.50 %	0.00 (0.52)	0.00 (0.52)	11.10 (19.26)	23.33 (28.84)	36.66 (37.21)	48.88 (44.36)	53.33 (46.91)	56.66 (48.83)	65.55 (54.06)	76.66 (61.15)	84.44 (66.79)	91.11 (72.72)	95.55 (77.99)	100.00 (89.47)	100.00 (89.47)	56.21 (49.21)
3	*P. nigrum* 0.75 %	0.00 (0.52)	0.00 (0.52)	14.44 (22.30)	50.00 (44.99)	56.66 (48.83)	60.00 (50.77)	68.88 (56.10)	71.10 (57.54)	80.00 (63.48)	85.55 (67.68)	93.33 (75.35)	96.66 (79.42)	100.00 (89.47)	100.00 (89.47)	100.00 (89.47)	65.10 (55.73)
4	*P. nigrum* 1.00 %	0.00 (0.52)	7.77 (16.11)	24.00 (29.58)	55.55 (48.19)	65.55 (54.08)	74.44 (59.64)	78.88 (62.66)	82.22 (65.08)	89.99 (71.72)	94.44 (76.51)	97.77 (82.80)	100.00 (89.47)	100.00 (89.47)	100.00 (89.47)	100.00 (89.47)	71.40 (61.55
5	*P. nigrum* 1.25 %	7.77 (16.11)	19.99 (26.51)	26.66 (30.97)	37.77 (37.92)	57.77 (49.49)	67.77 (55.64)	75.55 (60.37)	80.00 (63.48)	86.66 (68.68)	93.33 (75.35)	96.66 (81.32)	100.00 (89.47)	100.00 (89.47)	100.00 (89.47)	100.00 (89.47)	69.99 (61.58
6	*P. nigrum* 1.50 %	12.22 (20.42)	25.55 (30.23)	31.11 (33.89)	61.11 (51.42)	63.33 (52.78)	68.88 (56.14)	75.55 (60.41)	80.00 (63.60)	90.00 (71.72)	92.22 (73.87)	97.77 (82..80)	100.00 (89.47)	100.00 (89.47)	100.00 (89.47)	100.00 (89.47)	73.18 (63.68
7	*P. nigrum* 1.75 %	36.66 (37.25)	41.11 (39.86)	51.10 (45.63)	67.77 (55.45)	71.10 (57.58)	76.66 (61.22)	82.22 (65.08)	86.66 (68.57)	95.55 (77.99)	100.00 (89.47)	100.00 (89.47)	100.00 (89.47)	100.00 (89.47)	100.00 (89.47)	100.00 (89.47)	80.59 (69.70
8	*P. nigrum* 2.00 %	41.11 (39.87)	42.22 (40.52)	54.44 (47.60)	54.44 (53.49)	70.00 (56.99)	78.88 (62.86)	86.66 (68.68)	88.88 (70.57)	98.88 (86.14)	100.00 (89.47)	100.00 (89.47)	100.00 (89.47)	100.00 (89.47)	100.00 (89.47)	100.00 (89.47)	81.70 (70.90
9	Control	0.00 (0.52)	0.00 (0.52)	0.00 (0.52)	0.00 (0.52)	0.00 (0.52)	0.00 (0.52)	0.00 (0.52)	0.00 (0.52)	0.00 (0.52)	0.00 (0.52)	0.00 (0.52)	0.00 (0.52)	0.00 (0.52)	0.00 (0.52)	0.00 (0.52)	0.00 (0.52)
	Mean	10.86 (12.92)	15.18 (17.26)	24.07 (26.69)	41.72 (38.20)	49.38 (42.92)	57.28 (47.82)	63.57 (51.82)	66.41 (53.83)	74.19 (60.64)	79.38 (65.81)	83.94 (70.74)	86.54 (74.81)	87.77 (77.12)	88.64 (78.85)	88.88 (79.59)	

Each value is a mean of three replications
Values in parentheses are transformed arcsine values

	SED	CD (0.05)
Treatment	0.57	1.12 **
Period	0.73	1.45**
Treatment x period	2.21	4.36 **

1.50 per cent followed by *P. nigrum* 0.75 per cent (96.66%) and *P. nigrum* 0.50 per cent (91.11%) where as minimum mortality was observed in *P. nigrum* 0.25 per cent (91.10%). At 15 hr after the treatment cent percent mortality was observed in *P. nigrum* 0.25 per cent.

Similar results were reported by Ivbijaro and Agbage (1986) who reported that 0.1 g *P. nigrum* /20g of cowpea seed caused 100 per cent mortality of C. *maculatus* adults. In untreated control no mortality was observed up to 15 hr after the treatment. All the treatments were significantly superior to control.

P. nigrum 10 D (talc as a filler) (Pn10DT) was tested for its insecticidal action at 0.50, 1.00, 1.50, 2.00, 2.50, 3.00, 3.50, and 4.00 per cent in comparison with activated clay (1.00 %) and talc (4 %) (Table 2). Data on mortality of *C. maculatus* after 12, 24, 36, 48, 60, 72, 84, 96 and 108 hr the treatment were recorded. The cent per cent

Table 2 : Insecticidal action of *Piper nigrum* 10D (talc as filler) (Pn 10 DT) on *Callosobruchus maculatus*

S. No.	Treatments	% Adult mortality at									Mean
		12 hr	24 hr	36 hr	48 hr	60 hr	72 hr	84 hr	96 hr	108 hr	
1	Pn10DT 0.50 %	7.77 (16.11)	8.88 (17.27)	15.55 (23.19)	33.33 (35.24)	52.21 (46.27)	81.08 (64.36)	86.66 (68.68)	96.66 (81.32)	100.00 (89.47)	53.57 (49.10)
2	Pn10DT 1.00%	8.88 (17.27)	12.22 (20.42)	29.99 (32.87)	42.22 (40.52)	64.44 (53.51)	78.88 (62.75)	84.44 (66.86)	95.55 (79.85)	100.00 (89.47)	57.40 (51.50)
3	Pn10DT 1.50%	11.11 (19.42)	15.55 (23.19)	21.10 (27.28)	42.22 (40.51)	68.88 (56.12)	82.22 (65.15)	87.77 (69.67)	95.55 (79.85)	100.00 (89.47)	58.26 (52.30)
4	Pn10DT 2.00%	8.88 (17.27)	14.44 (22.30)	18.88 (25.74)	36.66 (37.25)	70.00 (56.81)	80.00 (63.48)	92.21 (74.36)	97.77 (84.66)	100.00 (89.47)	57.65 (52.37)
5	Pn10DT 2.50%	12.22 (20.42)	14.44 (22.30)	21.10 (27.24)	45.55 (42.44)	71.10 (57.54)	82.21 (65.19)	88.88 (70.83)	97.77 (84.66)	100.00 (89.47)	59.25 (53.34)
6	Pn10DT 3.00 %	14.44 (22.30)	16.66 (24.02)	21.11 (27.33)	46.66 (43.08)	70.00 (56.81)	77.77 (61.92)	87.77 (69.57)	94.44 (76.51)	100.00 (89.47)	58.76 (52.34)
7	Pn10DT 3.50 %	14.44 (22.30)	17.77 (24.91)	25.55 (30.35)	48.88 (44.36)	75.55 (60.53)	86.66 (68.68)	91.10 (73.20)	98.88 (86.14)	100.00 (89.47)	62.09 (55.55)
8	Pn10DT 4.00%	15.55 (23.19)	19.99 (26.51)	23.33 (28.84)	56.66 (48.83)	83.33 (66.19)	91.09 (73.46)	100.00 (89.47)	100.00 (89.47)	100.00 (89.47)	65.55 (59.49)
9	Activated clay 1.00%	92.22 (73.87)	100.00 (89.47)	100.00 (89.47)	100.00 (89.47)	100.00 (89.47)	100.00 (89.47)	100.00 (89.47)	100.00 (89.47)	100.00 (89.47)	99.13 (87.74)
10	Talc 4.00%	3.33 (10.51)	5.55 (13.47)	8.88 (17.27)	14.44 (22.30)	15.53 (23.18)	28.88 (32.45)	37.77 (37.92)	41.11 (39.87)	52.22 (46.27)	23.08 (27.03)
11	Control	0.00 (0.52)	3.33 (10.51)	5.55 (13.47)	8.88 (17.27)	11.11 (19.42)	15.55 (23.19)	17.67 (24.83)	18.88 (25.74)	24.44 (10.51)	11.71 (18.29)
	Mean	17.16 (22.11)	20.80 (26.76)	26.46 (31.19)	43.22 (41.93)	62.01 (53.26)	73.12 (60.92)	79.48 (66.81)	85.14 (74.32)	88.78 (80.10)	

mortality was seen in activated clay 1.00 per cent at 24 hr after the treatment followed by Pn10DT 4.00 per cent (19.99%), Pn10DT 3.50 per cent (17.17%) and Pn10DT 3.00 per cent (16.66%). At 60 hr after the treatment maximum mortality was seen in Pn10DT 4.00 per cent (83.33%) followed by Pn10DT 3.50 per cent (75.55%) and Pn10DT 2.50 per cent (71.10 %) against the control (11.11%) and talc 4 per cent (15.53%).

At 72 hr after the treatment Pn10DT 4.00 per cent registered 91.09 per cent mortality followed by Pn10DT 3.50 per cent (86.66%), Pn10DT 1.50 and 2.50 per cent (82.22%). All the beetles were dead in Pn10DT 4.00 per cent at 82 hr after the treatment while in Pn10DT 2.00, 3.50 and 2.50 per cent 92.21, 91.10 and 87.77 per cent beetles were dead respectively. At 96 hr after the treatment highest mortality was seen in Pn10DT 3.50 per cent (98.88%) followed by Pn10DT 2.00 and 2.50 per cent (97.77%) and Pn10DT 0.50per cent (96.66%). All the beetles were dead in Pn10DT 0.50, 1.00, 1.50, 2.00, 2.50, 3.00 and 3.50 per cent at 108 hr after the treatment. Juneja and Patel (1994) also reported that *P. nigrum* seed powder 5 parts /100 parts caused 100.00 per cent adult mortality to bruchids.

P. nigrum 10 D (fly ash as a filler) (Pn10DF) at same doses was used. Results (Table 3) revealed that maximum mortality was observed in activated clay 1.00 per cent (95.55%) followed by Pn10DF 4.00 per cent (45.55%), Pn10DF 3.50 per cent (33.33%) and Pn10DF 3.00 per cent (28.88%) at 12 hr after the treatment. At 24 hr cent per cent mortality was observed in activated clay 1.00 per cent followed by Pn10DF 4.00 per cent (59.99 %), Pn10DF 3.00 per cent (53.33%) and Pn10DF 2.50 per cent (49.99%).

At 36 hr after the treatment 66.66 per cent mortality was seen in Pn10DF 4.00 per cent which was followed by Pn10DF 3.00 per cent (61.11%) and Pn10DF 2.50 per cent (57.77%). The same trend was observed at 48 hr after the treatment. At 72 hr after the treatment cent percent mortality was registered in Pn10DF 4.00 per cent followed by Pn10DF 2.00, 2.50 and 3.00 per cent (88.88%) and Pn10DF 3.50 per cent (86.64 %). At 84 hr. after the treatment all the beetles were dead in Pn10DF 1.50. 2.00, 2.50, 3.00, and 3.50 per cent. At 96 hr after the treatment, in fly ash 4 per cent and untreated control 75.55 and 31.11 per cent mortality was recorded respectively. Miyakado *et al.*, (1979) observed that dust of *P. nigrum* was toxic to bruchids.

Table 3 : Insecticidal action of *Piper nigrum* 10D (fly ash as filler) (Pn 10 DF) on *Callosobruchus maculatus*

S. No.	Treatments	% Adult mortality at								Mean
		12 hr	24 hr	36 hr	48 hr	60 hr	72 hr	84 hr	96 hr	
1	Pn10DF 0.50 %	12.22 (20.42)	26.66 (30.90)	36.66 (37.25)	53.33 (46.91)	68.88 (56.10)	78.88 (62.66)	96.66 (81.32)	100.00 (89.47)	59.16 (53.13)
2	Pn10DF 1.00%	14.43 (22.29)	33.32 (35.15)	42.22 (40.52)	62.22 (52.07)	72.20 (58.24)	82.22 (65.08)	92.22 (73.87)	100.00 (89.47)	62.35 (54.59)
3	Pn10DF 1.50%	17.77 (24.91)	37.77 (37.89)	48.33 (44.04)	62.22 (52.09)	76.66 (61.15)	86.66 (69.01)	100.00 (89.47)	100.00 (89.47)	66.17 (58.51)
4	Pn 10 DF 2.00%	18.88 (25.74)	43.33 (41.15)	47.77 (43.72)	68.88 (56.10)	79.99 (63.48)	88.88 (70.57)	100.00 (89.47)	100.00 (89.47)	68.47 (60.00)
5	Pn10DF 2.50%	19.99 (26.51)	49.99 (44.99)	57.77 (49.47)	69.99 (56.85)	84.44 (66.86)	88.88 (70.57)	100.00 (89.47)	100.00 (89.47)	71.38 (61.78)
6	Pn10DF 3.00 %	28.88 (32.47)	53.33 (46.94)	61.11 (51.42)	72.21 (58.25)	83.33 (65.97)	88.88 (70.57)	100.00 (89.47)	100.00 (89.47)	73.46 (63.07)
7	Pn10DF 3.50 %	33.33 (35.24)	52.22 (46.28)	56.66 (48.83)	79.99 (63.48)	85.55 (67.68)	86.64 (68.56)	100.00 (89.47)	100.00 (89.47)	74.30 (63.63)
8	Pn10DF 4.00%	45.55 (42.43)	59.99 (50.80)	66.66 (54.75)	94.44 (79.17)	96.66 (83.50)	100.00 (89.47)	100.00 (89.47)	100.00 (89.47)	82.47 (72.38)
9	Activated clay 1.00%	95.55 (77.99)	100.00 (89.47)	100.00 (89.47)	100.00 (89.47)	100.00 (89.47)	100.00 (89.47)	100.00 (89.47)	100.00 (89.47)	99.44 (88.04)
10	Fly ash 4.00%	4.44 (11.99)	11.11 (19.42)	21.10 (27.28)	38.88 (38.56)	47.77 (43.72)	61.10 (51.44)	67.77 (55.41)	75.55 (60.37)	40.96 (38.52)
11	Control	0.00 (0.52)	3.33 (10.51)	6.66 (14.63)	13.33 (21.41)	15.55 (23.19)	21.11 (27.33)	28.88 (32.50)	31.11 (33.89)	14.99 (20.50)
	Mean	26.46 (29.14)	42.82 41.23	49.54 (45.58)	65.04 (55.85)	73.73 (61.76)	80.29 (66.82)	89.59 (79.04)	91.51 (81.78)	

Values in parentheses are transformed arcsine
i values mean of three replications

	SED	CD (0.05)
Treatment	0.81	1.61 **
Period	0.69	1.37**
Treatment x period	2.30	4.55 **

Black gram seeds treated with different concentrations of *P. nigrum viz.,* 0.25, 0.50, 0.75, 1.00, 1.25, 1.50, 1.75 and 2.00 per cent exhibited adverse effects on the insect (Table 4). In *P. nigrum* 2.00 per cent 6.6 eggs were laid which was significantly different from other treatments. Hatchability ranged between 68.33 and 85.23 per cent in different treatments. Minimum adult beetles emergence was from *P. nigrum* 1.50 per cent (1.66Nos). Among the different concentrations of *P. nigrum* seed powder, 0.25 per cent had maximum number of adults emerged from *P. nigrum* (16.00Nos).

Table 4 : Effect of *Piper nigrum* seed powder on oviposition, hatchability and adult emergence of Callosobruchus maculatus

S. No.	Treatments	No. of eggs laid	Hatchability %	No. of adults emerged*	% Seed weight loss after 45 days
1	*P. nigrum* 0.25%	27.66 (1.44)d	83.34 (66.16)b	16.00 (1.20)c	5.4 (13.42)c
2	*P. nigrum* 0.50%	13.66 (1.13)c	85.23 (67.42)c	7.33 (0.86)d	3.78 (11.17)bc
3	*P. nigrum* 0.75%	13.33 (1.17)c	78.57 (62.60)ab	5.33 (0.69)c	2.51 (8.81)b
4	*P. nigrum* 1.00%	12.00 (1.07)c	79.36 (55.71)a	4.00 (0.59)bc	2.06 (8.05)b
5	*P. nigrum*1.25%	11.66 (0.89)b	72.22 (58.24)a	4.66 (0.66)bc	2.85 (9.70)b
6	*P. nigrum* 1.50%	8.00 (0.89)b	68.33 (56.14)a	1.66 (0.20)a	0.53 (3.12)a
7	*P. nigrum* 1.75 %	11.33 (1.04)c	79.34 (62.67)ab	4.00 (0.59)bc	2.13 (8.29)b
8	*P. nigrum* 2.00%	6.66 (0.81)a	73.33 (57.89)a	2.66 (0.41)ab	0.78 (4.66)a
9	Control	246.00 (1.89)e	72.55 (58.42)a	124.00 (2.09)f	29.76 (33.05)d

Each value is a mean of 3 replication

* Figures in parentheses are transformed logarithmic values

**Figures in parentheses are transformed arcsine values

In a column means followed by same letter(s) are not significantly different (p=0.05) by DMRT

The results could be corroborated with the results of Olaifa and Erhun (1988) who reported that *P. guieneense* seed powder at low concentration significantly reduced the oviposition and adult emergence. Seed weight loss was minimum in *P. nigrum* 1.50 per cent (0.53%) followed by *P. nigrum* 2.00 per cent (0.78%) and these treatments were similar in their effect. Kumari and Singh (1998) who found that black pepper, *P. nigrum* powder was effective in reducing adult emergence and grain damage.

P nigrum 10D (talc as filler) (Pn10DT) at different concentrations *viz.,* 0.50, 1.00, 1.50, 2.00, 2.50, 3.00, 3.50 and 4.00 per cent was mixed with black gram seeds (Table 5). Minimum number of eggs were laid by female in Pn10DT 4.00 per cent (39.99 Nos) which was significantly different from other treatments. Aslam *et al.,* (2002) found that adult emergence and weight loss was less in chickpea treated with black pepper.

Table 5 : Effect of *Piper nigrum* 10 D (talc as filler) (Pn10DT) on oviposition, hatchability and adult emergence of *Callosobruchus maculatus*

S. No.	Treatments	No. of eggs laid	Hatcha-bility %	No. of adults emerged*	% Seed weight loss after 45 days	Seed germina-tion after 90 days**%
1	Pn10DT 0.50 %	58.66 (1.76)ef	75.38 (60.40)abc	34.33 (1.53)cd	7.40 (15.72)d	57.33 (49.22)e
2	Pn10DT 1.00 %	63.66 (1.80)fg	71.75 (57.90)ab	38.00 (1.58)d	6.93 (15.26)cd	61.33 (51.55)cd
3	Pn10DT 1.50 %	53.00 (1.72)de	71.15 (57.55)ab	32.33 (1.50)c	7.61 (16.01)d	65.33 (53.93)c
4	Pn10DT 2.00 %	50.66 (1.70)d	78.33 (62.45)bc	30.00 (1.47)bc	7.20 (15.49)cd	70.66 (57.22)c
5	Pn10DT 2.50 %	49.00 (1.69)cd	76.19 (60.79)abc	31.00 (1.49)c	7.15 (15.48)cd	80.00 (63.51)b
6	Pn10DT 3.00 %	45.66 (1.65)bc	71.04 (57.55)ab	26.33 (1.41)b	6.30 (14.52)c	80.00 (63.51)b
7	Pn10DT 3.50 %	43.3 (1.63)ab	73.97 (59.36)abc	26.00 (1.41)b	5.28 (13.28)b	82.66 (65.53)b
8	Pn10DT 4.00 %	39.66 (1.59)a	68.85 (56.09)a	18.33 (1.26)a	0.81 (5.18)a	90.66 (72.29)a
9	Activated clay 1.00%	65.00 (1.81)g	75.89 (60.63)abc	39.66 (1.59)d	7.75 (16.06)d	40.00 (39.61)f
10	Talc 4.00%	114.00 (2.05)h	80.65 (63.92)c	81.00 (1.90)e	12.00 (20.26)e	22.66 (28.41)g
11	Control	189.00 (2.27)i	89.75 (71.50)d	136.66 (2.13)f	17.75 (24.91)f	9.33 (17.71)h

Each value is a mean of 3 replications

*Values in parentheses are transformed logarithmic values

** Values in parentheses are transformed arcsine values

In a column means followed by same letter(s) are not significantly different (p=0.05) by DMRT

Hatchability in all the treatments was significantly lesser than untreated control. Pn10DT 4.00 per cent inhibited adult beetles emergence to the maximum level. Weight loss ranged between 0.18 (Pn10DT 4.00 per cent) and 7.40 per cent (Pn10DT 0.50 per cent). In untreated control 17.75 per cent weight loss was observed. Germination percentage was maximum in Pn10DT 4.00 per cent (90.66%) which was significantly different from other treatments.

Minimum number of eggs were laid by female in Pn10DF 4.00 per cent (22.66 Nos) while maximum number were laid in Pn10DF 0.50 per cent (61.33 Nos). Pn10DF 4.00 per cent was superior than other treatments (Table-6). Yahaya (2002) found ovicidal effect and reduced oviposition rate in *P. guieneense* treated cowpea seeds.

Table 6 : Effect of *Piper nigrum* 10 D (fly ash as filler) (Pn10DF) on oviposition, egg hatching and adult emergence of *Callosobruchus maculatus*

S. No.	Treatments	No. of eggs laid	Hatcha-bility %	No. of adults emerged*	% Seed weight loss after 45 days	Seed germina-tion after 90 days**%
1	Pn10DF 0.50 %	61. 33 (1.78)e	59.29 (50.36)b	28.00 (1.44)ef	12.03 (20.27)c	57.33 (49.22)c
2	Pn10DF 1.00 %	50.66 (1.70)d	73.73 (59.21)d	25.33 (1.40)de	11.11 (19.47)c	65.33 (53.93)cd
3	Pn10DF 1.50 %	50.33 (1.70)d	69.64 (56.19)bcd	22.33 (1.34)cd	6.11 (14.25)d	72.00 (58.09)c
4	Pn10DF 2.00 %	42.00 (1.62)c	57.12 (49.09)a	17.66 (1.24)c	4.78 (12.59)c	77.33 (61.59)b
5	Pn10DF 2.50 %	31.66 (1.50)b	63.15 (52.62)ab	13.33 (1.12)b	2.50 (8.98)c	80.00 (63.51)b
6	Pn10DF 3.00 %	30.00 (1.47)cb	64.49 (53.43)abc	11.66 (1.06)b	2.80 (9.59)b	81.33 (64.43)b
7	Pn10DF 3.50 %	28.00 (1.44)b	72.67 (58.51)cd	7.66 (0.87)a	0.35 (3.33)a	85.33 (67.52)b
8	Pn10DF 4.00 %	22.66 (1.35)a	73.44 (58.99)cd	7.00 (0.84)a	0.35 (3.33)a	93.33 (75.20)a
9	Activated clay 1.00%	41.33 (1.61)c	72.62 (58.47)e	20.33 (1.30)cd	6.43 (14.67)d	48.00 (43.85)e
10	Fly ash 4.00%	103.00 (2.01)f	84.80 (67.07)e	68.33 (1.48)g	16.18 (23.70)f	10.66 (18.98)g
11	Control	172.00 (2.23)g	93.22 (69.73)e	143.66 (2.15)f	19.31 (26.07)e	24.00 (29.28)f

Each value is a mean of 3 replications

*Values in parentheses are transformed logarithmic values

** Values in parentheses are transformed arcsine values

In a column means followed by same letter(s) are not significantly different (p=0.05) by DMRT

Ovipositional deterrent effect of *P. guineense* on *C. maculatus* was reported by Olaifa and Erhun.(1988). Hatchability was minimum in Pn10DF 2.00 per cent (57.12%) and maximum in Pn10DF 1.00 per cent (73.73%). In activated clay 1.00 per cent, fly ash 4 per cent and untreated control 72.62, 84.80 and 93.22 per cent hatchability was observed respectively. Adult emergence was minimum in Pn10DF 4.00 per cent (7.00Nos) followed by Pn10DF 3.50 per cent (7.66Nos) and these treatments were on par in their effect. Miyakado *et al.*, (1979) observed that dust of *P. nigrum* was toxic to bruchids. Loss of weight in seed was statistically comparable in Pn10DF 3.50 and 4.00 per cent. Umoetoock *et al.*, (2004) reported that *P. guineense* at 0.5g per 50g of cowpea seed registered less damage and weight loss due to *C. maculatus*.

References

Adams, J.M. and G.G.M. Schulten. 1978. Post harvest grain loss assessment methods. *Am. Assoc. Cereal Chem.*, St. Paul. Minnetota, U.S.A 193pp

Agarwal, A., S. Lal and K.C. Gupta, 1988. Natural Products as protectants of pulses against pulse beetle. *Bull. Grain Tech.*, 26: 154.

Aslam.M., Khalid. Alikhan and M,Z,H, Bajwa. 2002. Potency of some spices against *Callosobruchus chinensis* (Linnaeus). *J. Biol. Sci.*, 2(7): 449-452.

Babu, A., N. Raju, S. Albert, S. Ignacimuthu and S. Dorn. 1999.. Comparative efficacy of some indigenous plant extracts against the pulse beetle, *Callosobruchus maculatus* F. (Coleoptera: Bruchidae). *Biol. Agri. Res.,* and *Horticulture,* 17: 145-150

Caswell, G.H. 1981. The impact of infestation on commodities. *Tropical Stored Product Infestation,* 25: 19.

Credland, P.F.and A.W. Wright 1989. Factors affecting female fecundity in the cowpea seed beetle, *Callosobruchus maculatus* (Coleoptera : Bruchidae). *J. Stored Prod. Res.,* 25(3): 125-139.

Dongre, T.K., S.E. Pawar and M.R Harwallar. 1993. Resistance to *Callosobruchus maculatus* (Coleoptera: Bruchidae) in Pigeon pea (*Cajanus cajan* (L)) Millsp and other Cajanus species. *J. Stored Prod. Res.,* 29: 319-322

Hagstrum, D.W.1985. Pre harvest infestation of cowpeas by cowpea weevil (Coleoptera: Bruchidae) and population trends during storage in Florida. *J. Econ. Entomol.,* 78: 358-361

Ivbijaro, M. F and M. Agbaje. 1986. Insecticidal activities of *Piper guineense* Schum and Thonn on the control of *Callosobruchus maculatus. Insect Sci. Appl.,* 1(4): 521-524.

Jilani, G., M.M.I. Khan, and A.L. Ghiasuddin. 1998. Studies on insecticidal activity of some indigenous plant materials against the pulse weevil, *Callosobruchus analis* (F.). *Pak. J. Ent.,* 3:21 – 29.

Juneja, R.P. and J.R. Patel. 1994. Botanical materials as protectant of green gram, *Vigna radiata* (L.) Wilezek against pulse beetle, *Callosobruchus analis* Fabricous. *Gujarat Agric. Uni. Res..,* 20(1): 84-87.

Kumari, K. and S.N. Singh. 1998. Evaluation of efficacy of botanicals and insecticides against pulse beetle (*Callosobruchus chinensis* L.). *J. Appl. Biol.,* 8(1): 138-140.

Miyakado, M., I. Nakayama. H. Yoshioka and N. Nakatani. 1979. The Piperaceae amides I: structure of pipercide, a new insecticidal amide from *Piper nigrum* L. *Agri. Biol. Chem.,* 43 (7): 1609 – 1611.

Olaifa, J.I. and W.O. Erhun. 1988. Laboratory evaluation of *Piper guineense* for the protection of cowpea against *Callosobruchus maculatus. Insect Sci.& its Applic.,* 9(1): 55-59.

Umoetoock, S.P.A ., I.E. Oku and D.A. Ukeh. 2004. Reduction damage caused to stored cowpea (*Vigna unguiculata* L.) seeds by the bean bruchids (*Callosobruchus maculatus* F.) using plant products. *J. Food Technol.,* 2 (4) :195- 198.

Weaver, D.K., F.V. Dunkel .J.L Cusker and L.V. Puyvelde. 1992. Oviposition patterns in two species of bruchids (Coleoptera: Bruchidae) as influenced by the dried leaves of *Tetradenia riparia,* a perennial mint (Lamiales: Lamiaceae) that suppresses population size. *Environ. Entomol.,* 21(5): 1120-1121.

Yahaya, M.A. 2002. Effects of wood ash and dry fruit powder of *Piper guineense* on *Callosobruchus maculatus* (F). The Beam Sokoto. *J. Pure Appl. Sci.,* 6.

— xxx —

13

Evaluation of Repellent Property of Certain Native Botanical Powders Against Rice Weevil, *Sitophilus oryzae* L. (Curculionidae: Coleoptera) on Sorghum

SATHYASEELAN, V., A.M. AMALA HYACINTH AND T. SELVAMUTHUKUMARAN

*Lecturers, Department of Entomology, Faculty of Agriculture, Annamalai University, Annamalainagar – 608 002. Tamil Nadu.

SUMMARY

Results on repellent nature of certain native plant species revealed that Neem seed powder (77.4%) and persian lilac seed powder (75.6%) @ 15g. /100g. of sorghum seeds showed the highest repellent property 48 h after treatment followed by pongam seed powder (54.7%) and sweet flag rhizome powder (48.3%). The least repellency was recorded in the case of custard apple seed powder (23.3%).

Key words: Native Botanicals, Rice Weevil, Repellency

1. INTRODUCTION

Among the pests in sorghum, some are attacking both in the field and storage too. Rice weevil, *Sitophilus oryzae*. L. is one such major pest causing about 10.0 to 65.0 percent damage in stored seed. Under prolonged storage the damage hike up to 80.0 percent (Park, *et.al.*, 2003 and Louis *et. al.,2007)*. Sorghum being a consumable cereal, it is not advisable to adopt chemical control measures because of the hazardous residue problem of of the pesticides. However, scanty work has been done on the use of traditional plant products against the stored pest. The present study was therefore, undertaken to evaluate the repellent property of products of certain traditional plant species against rice weevil in sorghum.

2. MATERIALS AND METHODS

Products of plants namely, Neem (seed powder) (*Azadirachta indica* An.dr. Juss.), Pongam (seed powder) (*Pongamia glabra* L.), Sweet flag rhizome powder (*Acorus calamus* L.), Custard apple (seed powder) (*Annona squamosa* L.) and Persian lilac (seed powder L.) (*Melia azedarach* L.) at three dosages 5g, 10g, and 15g / 100g of sorghum seed were evaluated against the rice weevil.The present research was conducted under laboratory condition at the Department of Entomology, Faculty of Agriculture, Annamalai University during 2006- 2007.

2.1. Preparation of Native Botanical Powders

Seeds and rhizomes of five traditional plant species were collected and washed thoroughly, shade dried and ground to powder with a mixer grinder and later all the powders sieved by means of a 2mm sieve so as to obtain an uniform finely sized powder. This method is recommended to obtain a increased surface area of the traditional botanical powder, which in turn provide maximum coating over the surface of the sorghum seeds. (Tapondjou *et. al.*, 2002; Park *et. al.*, 2003 and Pendleton *et. al.*, 2005). The botanical powders were first mixed with acetone and then with the sorghum seeds thoroughly. A separate untreated control was also maintained with acetone alone.

Treated sorghum seeds were taken in petriplates and five pairs of one or two day old adult rice weevil were released into each petriplate, secured firmly with the rubber band. The set up was held under laboratory condition of temperature of 30±1°C and 70±5 percent relative humidity throughout the entire study.

Observations were recorded at 3, 6, 12, 24 and 48 hours after release of the insect. Effects of the treatments were assessed based on the orientation of the insect away from the treated sorghum seeds. The experiment was Completely Randomized Design with three replications in each treatment.

3. RESULTS AND DISCUSSION

The results of treatment of sorghum seeds with various herbal powders are presented in Table-1.

After three hours of insect release at a dose of @ 5g / 100g of sorghum seeds, neem seed powder repelled the pest by 95.3 per cent, followed by persian lilac seed powder with 91.5 per cent. Pongam seed powder exhibited a moderate repellency (61.3 per cent).

The level of repellency was maximum (98.5%) in neem seed powder followed by persian lilac seed powder (93.3%) while pongam seed powder (88.6%) exhibited a medium repellent effect against the weevil. Sweet flag rhizome powder and recorded lowest repellency.

At high dose of 15 g / 100g of sorghum seeds, neem seed powder showed maximum repellent activity of about 99.0 per cent when compared to other treatments. Followed my persian lilac seed powder 95.3% which was on par with that of pungam seed powder (90.5%) at the same interval of time. Sweet flag rhizome powder (69.8%) had a moderate repellency followed by the custard apple seed powder (53.7%) as reported by Kim *et. al.* (2003), Koul (2004), Tapondjou *et. al.* (2005) and Sahaf Bibi *et. al.* (2007).

After 6 hours of insect release, the repellency varied from 43.7 to 93.3 per cent followed by 45.6 to 96.7 per cent in the moderate dose and 48.2 to 98.5 per cent repellency was recorded in the higher dose of the herbal powders.

Among the treatments, the neem seed powder recorded the highest repellent effect of 93.3, 96.7 and 98.5 percent against the weevil followed by persian lilac seed powder and the pungam seed powder.

Among the treatments, the neem seed powder showed a highest repellent effect in all the three doses even after 12 hours of insect release. The repellency was reduced to 10 per cent in the case of neem seed powder followed by a 20 percent reduction in persian lilac and 25 per cent reduction in pongam seed powder. Similar trend was noticed in the remaining treatments also. Sweet flag rhizome powder recorded minimum repellency followed by custard apple powder (36.3, 39.4 and 41.8%). The present findings are in fune with that of Kouninki *et. al.* (2007).

Table 1 : Effect on the repellent property of certain native herbal powders against rice weevil

Treatments	Dose g / 100 g of seeds	% Repellency after				
		3 h	6 h	12 h	24 h	48 h
Neem seed powder	5	95.3ab (82.3)	93.3ab (80.7)	90.3ab (80.0)	86.0ab (78.0)	77.4b (65.4)
	10	98.5 a (89.7)	96.7a (82.9)	95.3a (82.3)	90.3a (80.0)	80.5ab (68.7)
	15	99.0a (89.7)	98.5a (89.7)	96.7a (82.9)	91.5a (80.1)	83.3a (70.2)
Pongam seed powder	5	86.7c (78.3)	80.5cd (68.7)	75.6d (63.8)	63.6ef (53.5)	54.6d (48.2)
	10	88.6bc (79.4)	83.3c (70.2)	78.2cd (67.3)	67.3e (55.1)	60.2cd (51.2)
	15	90.5b (80.0)	86.7bc (78.3)	81.5c (69.1)	70.6de (57.2)	63.6c (53.5)
Sweet flag rhizome powder	5	61.3e (51.8)	50.2ef (46.1)	48.2fgh (44.5)	45.1fg (42.0)	40.6h (40.1)
	10	64.7de (53.5)	53.7e (47.9)	51.3de (46.3)	48.2f (44.5)	43.1efg (41.0)
	15	69.8d (57.0)	58.2d (50.3)	53.7de (47.9)	50.2f (46.1)	45.7e (42.5)
Custard apple seed powder	5	48.2g (44.5)	43.7hi (41.2)	36.3hi (37.1)	31.7i (33.9)	24.6j (28.4)
	10	50.2fg (46.1)	45.6gh (42.4)	39.4h (39.5)	34.2hi (35.4)	28.5ij (32.0)
	15	53.7f (47.9)	48.2g (44.5)	41.8g (40.9)	37.3h (37.8)	31.3i (33.6)
Persian lilac seed powder	5	91.5b (80.1)	86.7bc (78.3)	81.5c (69.1)	76.6d (65.1)	75.6bc (63.8)
	10	93.3ab (80.7)	88.6bc (79.4)	83.3bc (70.2)	81.5c (69.1)	78.2b (67.3)
	15	95.3ab (82.3)	90.5ab (80.0)	86.7b (78.3)	83.3bc (70.2)	80.5ab (68.7)
	Control	0.0	0.0	0.0	0.0	0.0
	SE±	4.3	4.4	4.5	4.7	4.9
	C.D.(p=0.05)	8.7	8.8	9.1	9.5	9.9

Each value is a mean of three replications

Figures in parentheses are arc sine transformed values

Means followed by a common letter are not significant by DMRT (p=0.05)

After 48 hours of insect release, the reduction in the repellency was more than 25 per cent in neem seed powder followed by persian lilac seed powder and pungam seed powder treatment. Sweet flag rhizome powder and custard apple seed powder treatment, had 20 per cent reduction.

Among the treatments, neem seed powder exhibited a highest repellency effect (77.4, 80.5 and 83.3%) followed by persian lilac seed powder (75.6, 78.2 and 80.5%) against the weevil. Pungam seed powder showed a moderate repellency followed by sweet flag rhizome powder for the lower, medium and higher dose (40.6, 43.1 and 45.7%) respectively.

Thus results reveal that, as the doses increased with increase in the repellent effect vice versa as the duration increased. These results support findings of Lee *et. al.* (2004), Kordali *et. al.* (2006), Kavallieratos *et. al.* (2007) and Kestenholz *et. al.* 2007 against various stored pests.

It is inferred that among all the treatments, neem seed powder exhibited excellent repellency followed by persian lilac seed powder and pungam seed powder.

References

Kavallieratos, N.G. Athanassiou, C.G. Saitanis, C.J. Kontodimas, D.C. Roussos, A.N. Tsoutsa, M.S. Anastassopoulou, U.A. 2007. Effect of two azadirachtin formulations against adults of *Sitophilus oryzae* and *Tribolium confusum* on different grain commodities. *J. Food Prot.*, 70 (7): 1627-1632.

Kestenholz, C. Stevenson, P.C. Belmain, S.R. 2007. Comparative study of field and laboratory evaluations of the ethnobotanical, *Cassia sophera* L. (Leguminosae) for bioactivity against the storage pests, *Callosobruchus maculatus* (F.) (Coleoptera: Bruchidae) and *Sitophilus oryzae* (L.) (Coleoptera: Curculionidae). *J. Stored Prod. Res.*, 43(1): 79-86.

Kim, S.I. Roh, J.Y. Kim, D.H. Lee, H.S. Ahn, Y.J. 2003. Insecticidal activities of aromatic plant extracts and essential oils against *Sitophilus oryzae* and *Callosobruchus chinensis*. *J. Stored Prod. Res.*, 39(3): 293-303.

Kordali, S. Aslan, I. Calmasur, O. Cakir, A. 2006. Toxicity of essential oils isolated from three *Artemisia* species and some of their major components to granary weevil, *Sitophilus granarius* (L.) (Coleoptera: Curculionidae). *Industrial Crops and Products.*, 23(2): 162-170.

Koul, O. 2004. Biological activity of volatile di-n-propyl disulfide from seeds of Neem, *Azadirachta indica* (Meliaceae), to two species of stored grain pests, *Sitophilus oryzae* (L.) and *Tribolium castaneum* (Herbst). *J. Econ. Entomol,.* 97(3): 1142-1147.

Kouninki, H. Hance, T. Noudjou, F.A. Lognay, G. Malaisse, F. Ngassoum, M.B. Mapongmetsem, P.M. Ngamo, L.S.T. Haubruge, E. 2007. Toxicity of some terpenoids of essential oils of *Xylopia aethiopica* from Cameroon against *Sitophilus zeamais* Motschulsky. *J. Appl. Entomol.,* 131(4): 269-274.

Lee, B.H. Annis, P.C. Tumaalii, F. Choi, W.S. 2004. Fumigant toxicity of essential oils from the Myrtaceae family and 1,8-cineole against 3 major stored-grain insects. *J. Stored Prod. Res.,* 40(5): 553-564.

Louis, S. Delobel, B. Gressent, F. Duport, G. Diol, O. Rahioui, I. Charles, H. and Rahbe, Y.2007. Broad screening of the legume family for variability in seed insecticidal activities and for the occurrence of the A1b-like knottin peptide entomotoxins. Phytochemistry. *J. Stored Prod. Res.,* 68(4):521-535.

Park, I.K. Lee, S.G. Choi, D.H. Park, J.D. Ahn, Y.J. 2003. Insecticidal activities of constituents identified in the essential oil from leaves of *Chamaecyparis obtusa* against *Callosobruchus chinensis* (L.) and *Sitophilus oryzae* (L.). *J. Stored Prod. Res.,* 39(4): 375-384.

Pendleton, M.W. Vitha, S. Ellis, E.A. Chitio, F.M. Pendleton, B.B. 2005. Morphology of sorghum grain in relation to resistance to maize weevil. *International Sorghum and* Millets *Newsletter*. 46: 55-57.

Sahaf Bibi Zahra, Moharramipour, Saeid Meshkatalsadat and Mohammad Hadi. 2007. Chemical constituents and fumigant toxicity of essential oil from *Carum copticum* against two stored product beetles. *Insect Sci.,* (14)3: 213-218.

Tapondjou, A.L. Adler, C. Fontem, D.A. Bouda, H. Reichmuth, C. 2005. Bioactivities of cymol and essential oils of *Cupressus sempervirens* and *Eucalyptus saligna* against *Sitophilus zeamais* Motschulsky and *Tribolium confusum* du Val. *J. Stored Prod.* Res., 41(1): 91-102

Tapondjou, L.A. Adler, C. Bouda, H. Fontem, D.A. 2002. Efficacy of powder and essential oil from *Chenopodium ambrosioides* leaves as post-harvest grain protectants against six-stored product beetles. *J. Stored Prod. Res.,* 38(4): 395-402.

— xxx —

14

Management of Pulse Beetle, *Callosobruchus chinensis* Linn. in Chick Pea Through Botanicals

NANDIHALLI, B. S.
Department of Agrl. Entomology, College of Agriculture,
University of Agricultural Sciences, Dharwad - 580 005.

SUMMARY

Among different botanicals, sweet flag @ 1% was found effective in recording less number of pulse beetle eggs per 100 seeds and percentage of damaged seeds which was on par sweet flag @ 2% and boric acid @ 2%. Other plant products like dried neem leaf powder, neem cake, *Vitex negundo* L. leaf powder and *Vinca rosea* leaf powder were found inferior.

Key words : Bruchid, Botanicals, Internal Feeder.

1. INTRODUCTION

Chick pea is an important pulse in India and is the main source of protein for vegetarians. Unfortunately this pulse is prone to heavy damage by the pulse beetles, *Callosobruchus chinensis* L. in store. Bruchids breed rapidly in stored seeds and can complete a lifecycle in as little as 28 days. Adults live only for 10-12 days and females lay about 100 eggs. Even an extremely low bruchid population of undetectable levels of 2 bruchids/ton or 0.00001% of grains infested could result in major damage of 100% within 4–6 months. The bruchid causes weight loss, low the seed germination potential and quality deterioration. Sharma (1984) reported 11, 25 and 56 per cent damage in chick pea after 1, 3 and 6 months of infestation, respectively under controlled conditions.

Being an internal feeder, it is very difficult to control by normal methods. The prophylactic treatments are useful in minimizing storage losses due to the pest. The chemicals are hazardous to human beings and the practice of mixing insecticides with stored pulses is discouraged by legislation. This situation has awakend the need for less expensive and safer strategies among which natural products are potential. Vegetable oils and plant products have long been used in India for the protection of stored grains. Earlier workers reported the effectiveness of some of the plant powder such as neem seed powder, sweet flag, *Thevetia*, leaves of *Adatoda* and oils of neem, sunflower etc. against the pulse beetle (Pandey *et al.*, 1976; Khan, 1985 and Sibi, 2003). Wth this back drop, the present study was undertaken to screen certain plant products against the pulse beetle.

2. MATERIALS AND METHODS

An experiment was conducted at the College of Agriculture, University of Agricultural Sciences, Dharwad in 2004 under laboratory condition in completely randomized design with ten treatments (Table 1) and replicated three times. Annigeri 1 variety of chick pea was used in the experiment. In each treatment 250 g of chick pea was taken in a plastic container of 500 ml capacity. Muslin cloth was used for coveing the mouth of the container. As per the dosage, the plant products were mixed with chick pea seeds and then five pairs of pulse beetles were released for infestations into each container.

Observations were recorded on number of eggs per 100 seeds at 30,40, 60 and 90 days after imposition of treatments (DAI) and number of seeds damaged per 100 seeds at 40 60 and 90 DAI and data were subjected to statistical analysis.

3. RESULTS AND DISCUSSION

The data presented in table 1 indicate that among different botanicals tested, sweet flag @ 2% recorded significantly less number of eggs per 100 seeds (15.3 egg/100 seeds) which was on par with boric acid powder @ 2% (12.3 eggs/100 seeds), sweet flag @ 1% (20 eggs/ 100 seeds) and *Vinca rosea* @ 5% (20.7 eggs/100 seeds) at 30 days after imposing the treatments. Expectedly, the untreated control recorded significantly highest number of eggs (75 eggs/100 seeds). At 40 DAI, among botanicals sweet flag @ 2% was found significantly superior (17.7 eggs/100 seeds) which was on par with sweet flag @ 1% and *V. rosea* @ 5%. Boric acid powder @ 2% also behaved similarly. Two dosages of sweet flag tested were found on par with boric acid powder and these treatments were also found superior to other treatments at 60 and 90 DAI (Table 1).

Table 1 : Effect of botanicals on oviposition of pulse beetle, *Callosobruchus chinensis* in chick pea

Treatments	Number of eggs/100 seeds at			
	30 DAI	**40 DAI**	**60 DAI**	**90 DAI**
1. *Azadirachta indica* (leaves dry 5%)	22.70ab	38.3b	83.3c	260.0c
2. *Azadirachta indica* (cake 5%)	23.33ab	61.7cd	113.3d	350.0d
3. *Azadirachta indica* (cake 2%)	50.0b	50.0bc	76.7bc	260.0c
4. *Acorus calamus* 2%	15.3a	17.7a	20.0a	22.3a
5. *Acorus calamus* 1%	20.0a	20.3a	20.3a	27.0a
6. *Vitex negundo* 5%	46.0b	83.3de	115.0d	346.7d
7. *Vitex negundo* 2%	46.0b	86.7e	85.0c	350.0d
8. *Vinca rosea* 5%	20.7a	34.4ab	56.7b	83.3b
9. Boric acid powder 2%	12.3a	11.7a	8.0a	23.3a
10.Untreated control	75.0c	123.3f	120.0d	233.3c
CV%	14.73	15.18	21.82	20.58

DAI = Days after imposition of treatments

Regarding seeds damaged by the pulse beetle, sweet flag @ 2% recorded significantly less damaged seeds (3.00%) among botanicals at 40 DAI which was on par with *Vinca rosea* @ 5% (3.5%) and boric acid powder @ 2% (2.7%). The treatments like *V. negundo* leaf powder @ 5% (4.7%) and sweet flag @ 1% (5.3%) followed suit. However, at 60 and 90 DAI, two levels of sweet flag tested were found superior by recording less seed damage over other botanicals but were found on par with boric power @ 2% (Table 2). At 90 DAI, untreated control recorded significantly highest damaged seed (64.3%) which was on par with *V. negundo* leaf powder @ 2 and 5%.

Table 2 : Effect of botanicals on the seed damage of chick pea by pulse beetle, *Callosobruchus chinensis*

Treatments	Number of eggs/100 seeds at		
	30 DAI	40 DAI	60 DAI
1. *Azadirachta indica* (leaves dry 5%)	6.7cd	13.0b	36.0c
2. *Azadirachta indica* (cake 5%)	8.7de	18.3bc	52.3d
3. *Azadirachta indica* (cake 2%)	5.3bc	13.7bc	38.3c
4. *Acorus calamus* 2%	3.0ab	4.3a	4.0a
5. *Acorus calamus* 1%	5.3bc	5.3a	5.0a
6. *Vitex negundo* 5%	4.7bc	15.0bc	60.0de
7. *Vitex negundo* 2%	11.3ef	19.0c	63.3de
8. *Vinca rosea* 5%	3.5ab	13.3bc	19.3b
9. Boric acid powder 2%	2.7a	1.3a	1.7a
10.Untreated control	16.82g	18.3bc	64.3e
CV%	21.82	18.38	21.48

DAI = Days after imposition of treatments

Pandey *et al.* (1976) reported that sweet flag @ 1 to 2 per cent protected the chick pea seeds from damage by the bruchids which is in close agreement with the present findings. Long residual activity of sweet flag powder up to 90-120 days was reported against the bruchid in gram by Khan (1986) which is in line with the present findings. Eventhough boric acid powder @ 2% was found effective in reducing the egg laying and seed damage, it can not be recommended for the grains meant for consumption as it is carcinogenic in nature.

To conclude, sweet flag @ 1%, 2% can be recommended for the management of the pulse beetle in store house.

References

Khan,M.I, 1986, Efficacy of *Acorus calamus* L. rhizome powder against pulse beetle (*Callosobruchus chinensis*). *Punjabrao Krishi Research Journal,* 10: 72-74.

Pandey, N.D., Singh, S.R. and Tiwari, G.I., 1976, Use of some of plant powder, oil and extracts as protectants against pulse beetle, *Callosobruchus chinensis* Linn. *Ind. J. Ent.*, 38: 110-113.

Sharma, S.S., 1984, Review of literature on the loss caused by *Callosobruchus* species (Bruchidae: Coleoptera) during storage of pulses. *Bull. Grain Tech.*, 22(1): 62-71.

Sibi, V.G., 2003, Comparative biology and management of *Callosobruchus* spp. infesting soybean and other pulses with special reference to *Callosobruchus analis* Fabricius. M.Sc., (Ag). Thesis submitted to UAS, Dharwad, pp 110.

— xxx —

15

Evaluation of Botanical Formulations in the Management of Psocids in Paddy and Rice Storage

LOGANATHAN, M., K. SINGARAVADIVEL AND A. DAKSHINAMURTHY
Indian Institute of Crop Processing Technology, Thanjavur - 613 005 Tamil Nadu

SUMMARY

An attempt was made to evaluate the different botanical formulations *viz.*, Neem Seed Kernel Extract (NSKE-5%), Neem oil (NO - 1%) and Neem oil + pongam oil (NOPO - 1%) against grain lice. The results showed that the NSKE (5%) significantly reduced the psocids up to 52.16 and 34.17 per cent after 16 and 72 hrs of treatment respectively. An integrated management approach combining regular cleaning of storage premises and aeration along with application of NSKE (5%) twice in a week is more appropriate to manage this minute insect.

Key words: Psocid, Grain Storage, Botanical Formulations.

1. INTRODUCTION

The grain louse, *Liposcelis* spp. (Liposcelidae: Psocoptera) which is a minute with flattened body, causes nuisance to the personnel working in the grain storage premises (Sidik *et al.*, 1986) as they crawl all over the body of human beings and also the damage grain with loss in appearance and quality of commodities. Occurrence of the insect in the paddy and rice storage environment is well known in tropical countries (Rees and Walker, 1980). They also contributes to the contamination of food commodities as filth comprising exuviae, dead insects and their body fragments. The grains lice are aboundent in the paddy and rice storage godowns particularly in the coastal areas. The flattened body enables the insect to hide in the cracks and crevices of commodities, packaging materials and storage premises. Biology, act of problems and their control measures have been reviewed (Rajendran, 1994). Insecticides like, malathion, dichlorvos, deltamethrin and aluminium phosphide are normally used in the control of stored product insects. Among them aluminium phosphide was found to be effective against the psocid population (Loganathan and Singaravadivel, 2004). Regular application of insecticides for the management of insects in grain storage has led to development of resistance in insects and possibility of residues in the grains. The psocids can also be managed by other methods *viz.,* regular cleaning and aeration (Rajendran, 1994). Effect of botanical insecticides against storage pests has been studied in laboratory scale only. Hence, the present study, certain botanical insecticide formulations were evaluated for the management of the psocids in paddy and rice storage godowns.

2. MATERIALS AND METHODS

Certain botanical insecticide formulations, such as Neem oil (NO) and Neem oil + pongam oil (NOPO) were obtained from Tamil Nadu Agricultural University, Coimbatore. The Neem Seed Kernel Extract (NSKE-5%) was prepared by over night soaking of the powered neem seed kernel in water. After thorough mixing, the extract was filtered through cloth. The required concentration of NO (1%), NOPO (1%) and NSKE (5%) was prepared with water. Each formulation was sprayed in an area of 9 m^2 both on walls and floors of paddy and rice storage godown. An area of 225 cm^2 was marked for easy counting of psocids and five such places were marked in each treatment to record the observation. The observations on the psocid population prior to the treatment and after 16, 24, 48 and 72 hrs of the treatment were made in the marked area. The treatments were replicated five times. An untreated check was sprayed with water alone for comparison. The percent population reduction of the best due to the treatments over initial population was calculated and subjected to statistical analysis.

3. RESULTS AND DISCUSSION

The initial population of the psocid in the treatments ranged from 2366.43 to 2630.85 per square meter without any significant variation. The results showed that the neem seed kernel extract (5%) was found to reduce significantly the psocid population to 52.16, 35.61, 35.64 and 34.17 per cent significantly after 16, 24, 48 and 72 hrs of treatment respectively (Table 1). This was followed by NO (1%) and NOPO formulations (1%) in reducing the psocid population which ranged from 1 to 10 per cent only and there was no significant variation in the population compared to the untreated check.

Table 1 : Effect of botanical formulations against psocids in stored paddy and rice

Treatments	Number of psocids/m2 at				
	Pretreatment count	16 HAT	24 HAT	48 HAT	72 HAT
1. Neem Seed Kernel Extract	2466.42 a (NSKE - 5%)	1182.10 a (- 52.16)	1593.17 a (- 35.61)	1595.38 a (- 35.64)	1626.50 a (- 34.17)
2. Neem oil (NO - 1%)	2630.85 a	2559.74 b (- 2.71)	2604.18 c (- 1.01)	2355.32 b (- 10.49)	2408.65 b (- 8.45)
3. Neem oil + Pongam oil	2444.20 a (NOPO - 1%)	2337.54 b (- 4.36)	2310.88 b (- 5.45)	2266.44 b (- 7.27)	2310.88 b (- 5.45)
4. Untreated check	2366.43a	2466.42 b (4.23)	2488.64 b (5.17)	2488.64 b (5.17)	2608.63 b (10.24)

HAT - Hours After Treatment
* Average of 5 replications
Means followed by a common letter are not significantly different at 5% level by DMRT
Values in the parentheses are per cent reduction (-) or per cent increase (+) of psocid population over initial population.

Nayak *et a.l* (1998) reported that the phosphine was found to be highly effective against psocids. Loganathan and Singaravadivel (2004) reported that aluminium phosphide was found to be more effective (81%) up to 144 h of fumigation and malathion was least effective (11%) up to 48 of treatment in reducing the psocids in paddy and rice storage. The psocid population started increasing slowly after 48 of treatments in malathion applied grains. The present study revealed that the NSKE 5% was found to be more effective in reducing the psocids population. Earlier reports revealed that the azadirachtin was found to be effective in regulating the growth and metamorphosis in the stored product insect, *Corcyra cephalonica* (Chakraborti and Chatterjee, 2000). Since, there was no study to utilize botanical formulation to manage the psocids, our findings would demonstrate that the spray of NSKE 5% on the walls and floors of the paddy and rice storage godowns twice in a week in lieu of malathion, will be appropriate for the psocid menace.

The present study attempt to develop an integrated management approach along with regular cleaning of the storage premises, proper aeration and application of NSKE (5%) twice in a week appropriate to manage this minute insect in paddy and rice in storage.

References

Chakraborti, S and M.L Chatterjee. 2000. Effect of azadirachtin and some other neem pesticides on the development of rice moth, *Corcyra cephalonica* (Stainton). *J. Insect Sci.*, 13 (1 & 2):43- 48.

Loganathan. M. and K.Singaravadivel. 2004. Evaluation of insecticides for the control of grain lice in rice storage. *Pesticide Res J.*, 16(1) : 33-34.

Sidik, M, H. Halid and R.I. Pranata 1986. Pest problems and the use of insecticides in grain storage in Indonesia. pp. 37-43. In: Pesticides and humid tropical grain storage systems. Champ BR and Highley E (*Eds.*). ACIAR Proceedings No. 14.

Rees, D.P. and A.J. Walker 1990. The effect of temperature and relative humidity on population growth of three *Liposcelis* species (Psocoptera: Liposcelidae) infesting stored products in tropical countries. *Bull. Ent. Res.*, 80: 353 - 358.

Rajendran, S. 1994. Psocids in stored food commodities and their control. *Pestology*, 18 (4): 14 - 18.

Nayak, M.K, P.J. Collins and S.R. Reid. 1998 Efficacy of grain protectants and phosphine against *Liposcelis bostrychophila, L. entomophila* and *L. paeta* (Psocoptera: Liposcelidae). *Stored Prod. Quar. Ent.*, 91(5): 1208 - 1212.

— xxx —

16

Effect of Some Indigenous Pesticidal Plant Species Against Drug Store Beetle, *Stegobium paniceum* Linn. (Anobiidae : Coleoptera) in Coriander, *Coriandrum sativum* Linn.

KANNAN, R. AND V. SATHYASEELAN
Lecturers, Department of Entomology, Faculty of Agriculture, Annamalai University, Annamalai Nagar – 608 002.

SUMMARY

Efficacy of some indigenous plants like *Vitex negundo* L., *Aristolochia bracteolate* Lam., *Lantana camara* L., *Leucas aspera* (Wild) Linn., *Ocimum sanctum* L. and *Lawsonia inermis* L. was tested for their repellent property against drug store beetle *Stegobium paniceum* L.inn. in coriander seeds in laboratory. Results revealed that maximum repellency was registered by *V. negundo* leaf extract (63.33%), followed by *A. bracteolate* (56.7), *L. camara* (46.7), *L. aspera* (36.7%) and *O. sanctum* (30.33). Minimum effect was witnessed in coriander seeds treated dut to *L. inermis* 5% leaf extract (23.3%) at 72 hours after treatment. The repellency rate decreased with the increase in time.

Key words: Commodity, *Stegobium paniceum*, Repellency.

1. INTRODUCTION

Coriander, *Coriandrum sativum* Linnaeus is an important condiment, predominantly attacked by drug store beetle, *Stegobium paniceum* L. one of the widespread species of stored pests. When humidity of the food falls less than 6 per cent the larvae penetrate into the seeds and bore within it with a mild infestation. Several generations develop within and the insects do not come outside (Zakladnoi and Ratanova, 1987; Morgan *et al.,* 1998). Once they enter the seeds, the commodity will be completely destroyed rendering it unfit for consumption (Khare, 1993; Singh *et al.,* 2002). To curb this, the seeds have to be treated with chemicals but it is not feasible as it is a food commodity used in day today life. The use of synthetic organic chemicals though effective in controlling the pests, they leave residues hazards besides being costly. This makes us to find an alternative to do away the (José Iannacone *et al.,* 2005). Hence, leaf extracts of certain plant species were screened for protecting the coriander seeds against the store beetles in household conditions.

2. MATERIALS AND METHODS

An experiment was carried out at the Department of Entomology, Faculty of Agriculture, Annamalai University, during November 2006. Coriander seeds infested with *Stegobium paniceum* Linn. were used as mother culture and live culture was maintained. Coriander seeds were treated with leaf extract treatment:; *Vitex negundo* 5%, *Aristolochia bracteolata* 5%, *Lantana camara* 5%, *Leucas aspera* 5%, *Ocimum sanctum* 5%, *Lawsonia inermis* 5% and untreated control.

The leaf extracts were prepared with acetone 5%. The untreated check (T7) was maintained by mixing the seeds with acetone 5% alone. The coriander seeds were thoroughly mixed with the leaf extracts and shade dried.

Ten leaf extract treated coriander seeds were placed on a dry filter paper inside a glass petridish (70 mm). Five pairs of newly emerged adults of *S. paniceum* were released in each petriplate, secured firmly and kept in laboratory at ambient conditions with three replications per treatment.

Observations were recorded at 1, 3, 6, 12, 24, 48 and 72 hours after release. Efficacy of the leaf extracts was assessed based on the orientation of the insects away from the treated coriander seeds.

3. RESULTS AND DISCUSSION

Repellency of *S. paniceum* to different plant leaf extracts was assessed and the results are illustrated (Table-1).

Table 1 : Effect of certain indigenous pesticidal plant species against drug store beetle, *Stegobium paniceum* in coriander

Treatments	% Repellency (after the release of *S. paniceum*)						
	1 h	3 h	6 h	12 h	24 h	48 h	72 h
1. *Vitex negundo* 5%	15.33 (21.39)	23.33 a (28.86)	26.70 a (31.11)	36.70 a (37.29)	43.33 a (41.15)	60.33 a (50.94)	63.33 a (52.71)
2. *Aristolochia bracteolata* %	6.70 (15.00)	13.70 -ab (21.72)	20.00 ab (26.56)	23.33 ab (28.86)	36 70 ab (37.29)	50.00 a (45.00)	56.70 a (48.85)
3. *Lantana camara* 5%	3.33 (10.47)	6.70 b (15.00)	10.00 bc (18.44)	16.70 bc (24.12)	23.33 bc (28.86)	36.70 b (37.29)	46.70 b (43.11)
4. *Leucas aspera* 5%	3.33 (10.47)	6.70 b (15.00)	10.00 bc (18.44)	23.33 ab (28.86)	30.00 ab (33.21)	30.00 bc (33.21)	36.70 c (37.29)
5. *Ocimum sanctum* 5%	0.00 (0.91)	3.33 b (10.47)	3.33 bcd (10.47)	16.70 bc (24.12)	23.33 bc (28.86)	26.70 bcd (31.11)	30.00cd (33.21)
6. *Lawsonia inermis* 5%	3.33 (10.47)	6.70 b (15.00)	6.70 bcd (15.00)	10.00 bc (18.44)	13.70 cd (21.72)	20.00 cd (26.56)	23.33 d (28.86)
7. Untreated check	0 (0.91)	0 b (0.91)	3.33 bcd (10.47)	6.70 c (15.00)	6.70 d (15.00)	6.70 e (15.00)	6.70 e (15.00)
SE(d)	*NS*	6.68	5.93	6.31	6.48	5.74	3.73
CD (p=0.05)		14.32	12.73	13.53	13.91	12.31	8.01

Each value is a mean of three replications; Values in parentheses are arc sine transformed; Means followed by a ommon letter are not significantly different at 5% level by DMRT.

After 1h insect release, repellency of *V. negundo* leaf extract was minimum (13.33) but without significant difference between treatments. At 3 h of insect release, the repellency ranged between 3.3 and 23.33%. After 3 h, 23.33 % of insects were repelled from the *V. negundo* 5% leaf extract treated coriander seeds. The rest of the treatments behaved poorly revealing low repellency and are statistically on par with each other.

At 6 h only *V. negundo* leaf extract and *A. bracteolate* leaf extract had highest repellency 36.7 and 20.0 % respectively while other leaf extracts of *L. aspera, L. camara, O. sanctum* and *L. inermis* effected 20.0, 10.0, 10.0, 3.3 and 6.7% repellency respectively. After 24 h the repellency. *V.negundo* leaf extract range varied between 13.7 and

43.3%. The level of repellency was maximum (43.3%) in *V. negundo* treated coriander seeds followed by *A. bracteolata* treated seeds (36.7%) and *L. aspera* (30.0%). The other treatments registered <30% repellency and the minimum was in *L. inermis* treated seeds (13.7%). Statistically *V. negundo* is the only treatment showing significant difference among the treatments, whereas the other treatments are on par with each other concurring with the findings of Jilani and Saxena (1990) in *Lasioderma serricorne* and Xie *et al.* (1995) and Fields *et al.* (2001).

After 48 h of treatment, the repellency was >50% in *V. negundo* (60.3%) and *A. bracteolae* (50.0%) treated coriander seeds and followed by *L. camara* (36.7%) and *O. sanctum* (30.0%). The repellency rate was similar as that in the 24 h observation and the minimum was in *L. inermis* treated coriander seeds (20.0%). Even though high repellency was registered in *V. negundo* and *A. bracteolata* treated coriander seeds (60.3 and 50.0% repellency) statistically both of them were alike.

After 3 days of treatment, *V. negundo* leaf extract 5% and *A. bracteolata* leaf extract 5% recorded the higher order of repellency (63.3 and 56.7% respectively) and increased the movement of the beetle out of the seeds and this observation is similar to the reports of Mohan and Fields (2002) in *Tribolium castaneum*, followed by *L. camara* and *L. aspera*. Similar to all the above findings *L. inermis* exhibited minimum repellency (23.3%) even after 3 days after the insect release.

The laboratory results revealed that the leaf extracts of *V.negundo* and *A. bracteolata* had repelled *S. paniceum* (more than 50%) after 48 h of treatment and similar results were obtained by Jilani *et al.* (2001) in *T.castaneum* and Malik and Mujtabe-Naqvi (1984) in four different storage insects.

From the study, rate of repellency was higher in *V. negundo* leaf extract than by *A. bracteolata* leaf extract, whereas in the other treatments even after 72 h of the insect release the repellency did not exceed 50% and *L. inermis* had least repellency.

References

Jilani G, Saxena RC, Rueda BP. 2001. Repellent and growth-inhibiting effects of Turmenic oil, Sweetflag oil, Neem oil, and "Margosan-O" on Red flour beetle (Coleoptera: Tenebrionidae). *J. Econ. Entomol.*, 81 : 1226-30.

José Iannacone, Hildebrando Ayala and Amid Román 2005 Toxicological effects of four plants on corn borer *Sitophilus zeamais* motschulsky 1855 (Coleoptera:

Curculionidae) and on drugstore beetle, *Stegobium paniceum* (Mlinnaeus 1761) (Coleoptera: anobiidae) in Peru. *Gayana*, 69(2) : 234-240.

Khare B.P. 1993. Stored grain pests and their management. Kalyani Publishers, Ludhiana, 314p.

Kim SI, Park C, Ohh MH, Cho HC, Ahn YJ. 2003. Contact and Fumigant Activities of Aromatic Plant Extracts and Essential Oils Against *Lasioderma serricorne* (Coleoptera : Anobiidae). *J. Stored Prod. Res.*, 9(1) : 11-9.

Malik M, Mujtabe-Naqvi SH. 1984. Screening of some indigenous plants as repellents or antifeedants for stored grain insects [*Tribolium castaneum, Rhyzopertha dominica, Saussurea lappa, Chenopodium ambrosioides*]. *J. Stored Prod. Res.*, 20 : 41- 44.

Mohan S, Fields PG. 2002. A Simple Technique to Assess compounds that are repellent or attractive to Stored-Product Insects. *J. Stored Prod. Res.*, 38(1) : 23-31.

Morgan C, Sherington J, Gudrups I, Bowden NS. 1998. The Assessment of Potential Attractants to Beetle Pests: Improvements to Laboratory Pitfall Bioassay Methods. *J. Stored Prod. Res.*, 34(1) : 59-74.

Singh, G., I.P. Kapoor, S.K. Pander, U.K. Singh & R.K. Singh. 2002. Studies on essential oils. Part 10; antibacterial activity of volatile oils of some spices. *Phytother. Res.*, 16: 680-682

Su HCF. 1986. Laboratory evaluation of the toxicity and repellency of coriander seed to four species of stored-product insects. *J. Entomol. Sci.*, 21 : 169-74.

Xie YS, Fields PG, Isman MB. 1995. Repellency and toxicity of azadirachtin and neem to three stored-product insects. *J. Econ. Entomol.*, 88 : 1024-31.

Zakladnoi G.A. and Ratanova V.F. 1987. Stored grain pests and their control. Oxanian Press Pvt Ltd., New Delhi, 268 p.

— xxx —

17

Effect of Leaf Dusts of Certain Plant Species Against Pulse Beetle, *Callosobruchus chinensis* Linn. in Black Gram

MAHENDRAN, M.,[+] B. RAVIVARMAN[+] AND C. KATHIRVELU*

[+]PG Students, *Lecturer, Department of Entomology, Faculty of Agriculture, Annamalai University, Annamalai Nagar – 608 002

SUMMARY

Screening of twenty plant species in the form of leaf dust @ 3% and 5% dosages against pulse beetles on black gram, revealed that maximum mean mortalities of the adult beetles of 82.38 and 89.76 % were recorded in the seeds treated with marigold (*Tagetes erecta*) leaf dust @ 3% and 5% dosages respectively, whereas the wormwood (*Artemisia vulgaris*) treatment caused 81.19 and 83.56 % mortality @ 3% and 5% respectively, followed by henna (*Lawsonia inermis*) leaf dust with 75.79 and 81.90 % mortality at 3% and 5% dosages respectively.

Key words: Leaf Dust, Pulse Beetle, Black Gram

1. INTRODUCTION

The pulse beetle, *Callosobruchus chinensis* Linn. (Coleoptera: Bruchidae) is a major pest of leguminous grains such as black gram, cowpea, and green gram in storage, (Raja *et al.* 2000; Park *et al.* 2003). It causes substantial loss to grains during storage due to the pest infestation either carried over from the field or occurs in storage. The larvae bore into the pulse grains, feed and develop inside, which become unsuitable for human consumption and germination is also affected. The plant-derived materials are more readily biodegradable. Some are less toxic to mammals, may be more selective in action, and may retard the development of resistance. Their main advantage is that they may be easily and cheaply produced by the farmers. Screening of the plants in order to develop new botanical insecticides as alternatives to the existing insecticides is therefore required. The present study was hence, carried out to determine the effect of certain plant species as leaf dusts against of pulse beetle, *C. chinensis* in black gram.

2. MATERIALS AND METHODS

A laboratory experiment was conducted at the Department of Entomology, Faculty of Agriculture, Annamalai University, during 2007. *C. chinensis* was reared on black gram seeds in glass jars covered with muslin cloth by following the method developed by Credland and Wright (1989). Leaves of the following plants species were collected, shade dried, powdered using a pulverizer and used in the study.

Sl.No	Common name	Botanical name	Family
1	False daisy	*Eclipta alba L.*	Amaranthaceae
2	Rough chaff	*Achyranthes aspera* L.	Amaranthaceae
3	Marigold	*Tagetes erecta* L.	Asteraceae
4	Fish poison	*Tephrosia purpurea* L.	Fabaceae
5	Asian pigeonwings	*Clitoria ternatea* L.	Fabaceae
6	Sesbania	*Sesbania grandiflora* L.	Fabaceae
7	Holy basil	*Ocimum sanctum* L.	Lamiaceae
8	Curry leaf	*Murraya koenigii* L.	Rutaceae
9	Black nightshade	*Solanum nigrum* L.	Solanaceae
10	Tobacco	*Nicotiana tabacum* L.	Solanaceae
11	Indian privet	*Vitex negundo* L.	Verbenaceae
12	Jatropha	*Jatropha curcas* L.	Euphorbiaceae
13	Wormwood	*Artemisia vulgaris* L.	Asteraceae
14	Bael	*Aegle marmelos* L.	Asparanjanaceae
15	Oleander	*Nerium oleander* L.	Apocyanaeceae
16	Periwinkle	*Catharanthus roseus* L.	Lamiaceae
17	Mint	*Mentha spicata* L.	Lamiaceae
8	Eucalyptus	*Eucalyptus globulus* Latrill	Myrtaceae
19	Henna	*Lawsonia inermis* L.	Lithraeceae
20	Sweet basil	*Ocimum basilicum* L.	Lamiaceae

Twenty grams of black gram seeds at 12 per cent moisture content were taken in petridishes. The leaf powders 3% and 5% (w/w) were added separately to black gram seeds and shaken thoroughly. Twenty newly emerged adults were released into each petridish and kept in the laboratory. Mortality was recorded at 24 h intervals for seven days.The experiment was laid in RBD with three replications.

3. RESULTS AND DISCUSSION

Maximum mortality of Bruchus was observed in 1, 2 and 3 days after treatment with *T. erecta* (40.00, 53.33 and 83.33 %) followed by *A. vulgaris* (28.33, 60.00 and 80.00 %) and *L. inermis* (28.33, 48.33 and 65.00 %) (Table-1). Among the plant powders, cent per cent mortality was registered in *T. erecta* and *A. vulgaris* followed by *L. inermis* (86.66 %) after 4 days of the treatment. The results are in confirmation with the earlier works of Singh (2003), who found that the leaf powders effectively protected the grains of khesari from *C. chinensis*. All the beetles were dead in *L. inermis* at 5 days after treatment while in *J. curcas, A. marmelos and Catharanthus,* the mortality was 86.66, 86.66 and 83.33 per cent respectively. At 6 and 7 days after treatment maximum mortality was seen in *A. marmelos* (96.65 and 98.33 %) and *Catharanthus* (88.33 and 98.33 %) followed by *J.curcus* (96.66 and 96.66 %) whereas minimum mortality was observed in *S. grandiflora* (61.66 and 68.33%).

The results revealed that maximum mortality was observed in1 and 2 days after treatment with *T. erecta* (50.00 and 78.33 %) followed by *A. vulgaris* (36.66 and 56.66 %) and *L. inermis* (33.33 and 56.66 %) while in control all the insects were alive (Table -2). Maximum mortality was observed in *T. erecta* (83.33%) followed by *A. vulgaris* (91.66 %) and *L. inermis* (83.33 %) 3 days after treatment. Singh *et al.,* (2001) reported the similar findings effectively checking the population and growth of *C. chinensis*. After 4 days of the treatment cent per cent mortality was observed in *A. vulgaris* and *L. inermis* followed by *O. canum* and A. *marmelos* (73.33 %). At 5 and 6 days after treatment maximum mortality was seen in *E. globulus* (85.00 and 96.66 %).

A.*marmelos* (83.33 and 93.33%) and *N.tabacum* (81.66 and 86.66%) followed by *O.canum* (80.00 and 86.66%), mint (76.66 and 90.00%)and *J. curcus* (76.66 and 83.33%). These results concur with Saxena and Sukumaran (2005), who found that, tobacco powder reduced population of *C. maculates* (80.71 %). After 7 days of the treatment,

Table 1 : Effect of plant leaf 3 % dusts of certain plant species against *Callosobruchus chinensis* in black gram

Sl. No	Treatments	Per cent mortality at							
		1 DAT	2DAT	3DAT	4DAT	5DAT	6DAT	7DAT	Mean
1	*Eclipta alba*	3.33 (9.56)	18.33 (25.19)	30.00 (33.16)	43.33 (41.16)	56.66 (48.83)	70.00 (56.84)	75.00 (60.07)	42.37 (39.26)efg
2	*Achyranthes aspera*	1.66 (6.21)	21.66 (27.71)	33.33 (35.25)	46.66 (43.08)	61.66 (51.75)	73.33 (58.93)	81.66 (64.69)	45.70 (41.09)ef
3	*Tagetes erecta*	40.00 (39.21)	53.33 (46.92)	83.33 (66.26)	100.00 (87.13)	100.00 (87.13)	100.00 (87.13)	100.00 (87.13)	82.38 (71.56)a
4	*Tephrosia purpurea*	3.33 (9.56)	16.66 (46.92)	26.66 (31.07)	38.33 (38.24)	46.66 (43.08)	61.66 (51.75)	71.66 (57.86)	37.85 (36.51)h
5	*Clitoria ternatea*	1.66 (6.21)	13.33 (21.14)	28.33 (32.09)	46.66 (43.08)	56.66 (48.83)	75.00 (60.07)	83.33 (66.14)	43.56 (39.65)fg
6	*Sesbania grandiflora*	0.00 (2.86)	6.66 (14.75)	18.33 (25.30)	30.00 (33.16)	41.66 (40.19)	61.66 (51.75)	68.33 (55.85)	32.37 (31.98)d
7	*Ocimum canum*	10.00 (18.04)	21.66 (27.71)	40.00 (39.21)	61.66 (51.75)	76.66 (61.33)	91.66 (75.30)	95.00 (78.59)	56.66 (50.27)l
8	*Murraya koenigii*	3.33 (9.56)	13.33 (21.33)	25.00 (29.92)	40.00 (39.21)	50.00 (45.00)	68.33 (55.82)	83.33 (66.14)	40.47 (38.14)fgh
9	*Solanum nigrum*	11.66 (19.88)	18.33 (25.19)	38.33 (38.24)	48.33 (44.04)	61.66 (51.75)	71.66 (57.86)	78.33 (62.29)	46.9 (42.75)e
10	*Nicotiana tabacum*	1.66 (6.21)	15.00 (22.78)	31.66 (34.23)	50.00 (45.00)	63.33 (52.74)	80.00 (63.55)	91.66 (73.40)	47.61 (42.56)e
11	*Vitex negundo*	1.66 (6.21)	18.33 (25.19)	35.00 (36.23)	43.33 (41.15)	51.66 (45.95)	61.66 (51.75)	71.66 (57.91)	40.47 (37.70)fgh
12	*Jatropa curcas*	11.66 (19.88)	31.66 (34.23)	50.00 (45.00)	65.00 (53.76)	86.66 (68.85)	96.66 (81.94)	96.66 (81.94)	62.61 (55.08)c
13	*Artemisia vulgaris*	28.33 (32.09)	60.00 (50.79)	80.00 (63.55)	100.00 (87.13)	100.00 (87.13)	100.00 (87.13)	100.00 (87.13)	81.19 (70.71)a
14	*Aegle marmelos*	15.00 (22.59)	36.66 (37.25)	51.66 (45.95)	75.00 (60.07)	86.66 (69.24)	96.66 (81.94)	98.33 (83.78)	65.71 (57.26)c
15	*Nerium oleander*	3.33 (9.56)	13.33 (21.33)	25.00 (29.92)	38.33 (38.24)	48.33 (44.04)	63.33 (52.74)	78.33 (62.29)	38.56 (36.87)gh
16	*Catharanthus roseus*	15.00 (22.59)	33.33 (35.25)	61.66 (51.75)	76.66 (61.14)	83.33 (66.14)	88.33 (70.11)	98.33 (83.78)	65.23 (55.82)c
17	*Mentha spicata*	6.66 (14.75)	16.66 (24.04)	30.00 (33.16)	53.33 (46.92)	66.66 (54.83)	75.00 (60.07)	86.66 (68.66)	47.85 (43.20)e
18	*Eucalyptus globulus*	5.00 (11.40)	21.66 (27.71)	38.33 (38.24)	50.00 (45.00)	56.66 (48.84)	68.33 (56.84)	78.33 (62.29)	45.47 (41.47)ef
19	*Lawsonia inermis*	28.33 (31.07)	48.33 (44.04)	65.00 (53.84)	86.66 (68.66)	100.00 (87.13)	100.00 (87.13)	100.00 (87.13)	75.47 (65.57)b
20	*Ocimum bacilicum*	6.66 (14.75)	20.00 (26.45)	33.33 (35.25)	46.66 (43.08)	58.33 (49.83)	68.33 (55.85)	76.66 (61.14)	44.28 (40.91)ef
21	Untreated control	0.00 (2.86)	0.00 (2.86)	0.00 (2.86)	3.33 (9.56)	5.00 (11.40)	8.33 (16.59)	11.66 (19.88)	4.04 (9.43)m

Figures in parentheses are transformed arcsine values
Mean values followed by common letter are not significantly different at 5 % level (DMRT)
Each value is a mean of 3 replications.

Table 2 : Effect of leaf 5 % dusts of certain plant species against *Callosobruchus chinensis* in black gram

Sl. No	Treatments	Per cent mortality at							
		1 DAT	2DAT	3DAT	4DAT	5DAT	6DAT	7DAT	Mean
1	*Eclipta alba*	6.66 (14.75)	18.33 (25.19)	36.66 (37.22)	53.33 (46.91)	63.33 (52.77)	81.66 (64.69)	83.33 (65.95)	49.04 (43.93)j
2	*Achyranthes aspera*	11.66 (19.88)	28.33 (33.16)	43.33 (41.15)	56.66 (48.86)	76.66 (61.33)	81.66 (64.69)	86.66 (68.66)	54.99 (48.25)gh
3	*Tagetes erecta*	50.00 (44.99)	78.33 (62.48)	100.00 (87.13)	100.00 (87.13)	100.00 (87.13)	100.00 (87.13)	100.00 (87.13)	89.76 (77.59)a
4	*Tephrosia purpurea*	3.33 (8.05)	13.33 (21.33)	28.33 (32.14)	41.66 (40.19)	51.66 (45.95)	71.66 (57.86)	78.33 (62.40)	41.18 (38.27)k
5	*Clitoria ternatea*	8.33 (18.04)	18.33 (25.00)	31.66 (35.21)	46.66 (43.08)	61.66 (51.75)	73.33 (62.29)	86.66 (68.66)	47.37 (43.43)j
6	*Sesbania grandiflora*	8.33 (18.04)	23.33 (28.85)	36.66 (37.20)	48.33 (44.04)	60.00 (50.79)	68.33 (55.77)	81.66 (64.69)	64.99 (55.85)d
7	*Ocimum canum*	16.66 (23.74)	36.66 (37.20)	61.66 (51.75)	73.33 (58.93)	80.00 (63.55)	86.66 (68.66)	100.00 (87.13)	46.66 (42.77)j
8	*Murraya koenigii*	3.33 (11.40)	23.33 (28.78)	38.33 (38.22)	53.33 (46.92)	66.66 (54.78)	81.66 (64.69)	100.00 (87.13)	52.37 (47.42)hi
9	*Solanum nigrum*	8.33 (16.21)	26.66 (30.94)	48.33 (44.04)	60.00 (50.79)	65.00 (53.76)	71.66 (57.86)	80.00 (63.55)	51.42 (45.30)ij
10	*Nicotiana tabacum*	10.00 (17.46)	28.33 (32.14)	53.33 (47.88)	68.33 (55.85)	81.66 (64.69)	86.66 (68.66)	100.00 (87.13)	61.18 (53.40)de
11	*Vitex negundo*	8.33 (14.69)	21.66 (27.59)	38.33 (38.19)	55.00 (48.86)	63.33 (52.77)	73.33 (58.93)	81.66 (64.69)	48.80 (43.67)j
12	*Jatropa curcas*	16.66 (24.04)	38.33 (38.22)	53.33 (46.91)	63.33 (52.74)	76.66 (61.14)	83.33 (68.66)	100.00 (87.13)	61.66 (54.12)d
13	*Artemisia vulgaris*	36.66 (37.20)	56.66 (48.86)	91.66 (73.40)	100.00 (87.13)	100.00 (87.13)	100.00 (87.13)	100.00 (87.13)	83.56 (72.57)b
14	*Aegle marmelos*	28.33 (32.09)	46.66 (43.08)	60.00 (50.79)	73.33 (58.93)	83.33 (65.95)	93.33 (76.75)	100.00 (87.13)	69.28 (59.25)c
15	*Neerium oleander*	8.33 (18.04)	21.66 (28.85)	33.33 (36.23)	43.33 (46.13)	58.33 (49.83)	68.33 (55.85)	83.33 (65.95)	45.23 (42.98)j
16	*Catharanthus roseus*	23.33 (28.85)	36.66 (37.25)	48.33 (44.04)	58.33 (49.80)	76.66 (61.33)	80.00 (63.55)	86.66 (68.66)	58.56 (50.50)fg
17	*Mentha spicata*	16.66 (24.04)	26.66 (32.09)	40.00 (39.21)	56.66 (46.92)	76.66 (61.14)	90.00 (71.95)	96.66 (81.94)	57.61 (51.04)ef
18	*Eucalyptus globulus*	10.00 (18.04)	31.66 (34.14)	53.33 (46.92)	70.00 (57.98)	85.00 (67.40)	96.66 (81.94)	96.66 (81.94)	63.33 (55.48)d
19	*Lawsonia inermis*	33.33 (36.23)	56.66 (48.86)	83.33 (67.40)	100.00 (87.13)	100.00 (87.13)	100.00 (87.13)	100.00 (87.13)	81.90 (71.57)b
20	*Ocimum bacilicum*	21.66 (27.59)	33.33 (35.16)	48.33 (44.03)	63.33 (53.76)	73.33 (59.05)	83.33 (65.95)	91.66 (73.40)	59.28 (51.28)ef
21	Untreated control	0.00 (2.86)	0.00 (2.86)	0.00 (2.86)	0.00 (2.86)	5.00 (11.40)	5.00 (11.40)	10.00 (18.43)	2.85 (7.53)l

Figures in parentheses are transformed arcsine values

Mean values followed by common letter are not significantly different at 5 % level (DMRT)

Each value is a mean of 3 replications.

	SED	CD (0.05)
Treatment	1.17	2.31 **
Days	0.67	1.33 **
Treatment x Days	3.11	6.13. **

cent per cent mortality was recorded with A. *marmelos, O. canum, N. tabacum, M. koenigii, J. curcus* followed by *M. spicata* and *E. globulus* (96.66 %) whereas lowest mortality was observed in *T. purpurea* (78.33 %).

It is concluded that maximum mean mortality of the adult beetles 82.38 and 89.76 % was recorded in the seeds due to marigold (*T. erecta*) leaf dust @ 3% and 5% dosages respectively followed by the wormwood (*A. vulgaris*) treatment had caused 81.19 and 83.56 % mortality at 3% and 5% dose respectively, Henna (*L. inermis*) leaf dust with 75.79 and 81.90 % mortality at 3% and 5% concentrations respectively. Among the plant leaf dusts tested against pulse beetle, Marigold (*T. erecta*), wormwood (*A. vulgaris*) and henna (*L. inermis*) checked the pulse beetle population effectively in black gram. Hence, these plants require further studies to confirm the molecules responsible for their action for long storage of the grains.

References

Credland, P.F. and A.W. Wright. 1989. Factors affecting female fecundity in the cowpea seed beetle, *Callosobruchus maculatus* (Coleoptera: Bruchidae). *J. Stored Prod. Res.,* 25(3): 125-139.

Park, C., S.I. Kim and Y.J Ahn. 2003. Insecticidal activity of asarones identified in *Acorus gramineus* rhizome against three coleopteran stored-product insects. *J. Stored Prod. Res.,* 39(3):333–342.

Raja, N., S. Albert. and S. Ignacimuthu. 2000. Effect of solvent residues of *Vitex negundo* Linn. and *Cassia fistula* Linn. on pulse beetle and its larval parasitoid, *Dinarmus vagabundus* (Timberlake). *Ind. J. Expt. Biol.,* 38(3): 27-33.

Saxena, R. and S.K. Sukumaran. 2005. Efficacy of tobacco leaf powder (*Nicotiana tobaccum*) as seed protectant against *C. maculatus* Fab. (Coleoptera: Bruchidae). *Insect Envt.,* 11(2): 65-66.

Singh, R., B. Singh and R. A. Verma. 2001. Efficacy of different indigenous plant powders as grain protectants against *Callosobruchus chinensis* Linn. on pea. *Ind. J. Ent.,* 63(2): 179-181.

Singh, S.C. 2003. Effect of neem leaf powder on infestation of the pulse beetle, *Callosobruchus chinensis* in stored khesari. *Ind. J. Ent.,* 65(2): 188-192.

— xxx —

18

Effect of Milkweed Plant, *Calotropis gigantea* R. Br Powders Against Rice Weevil (*Sitophilus oryzae* L.) on Paddy

RANI, T. AND S. PUSHPALATHA
*Lecturers, Department of Entomology, Faculty of Agriculture, Annamalai University, Annamalainagar - 608 002.

SUMMARY

Results of evaluation of milkweed plant in the form of leaf powder and flower powder against rice weevil @ three concentrations viz., 1, 3 and 5 per cent in each revealed that, 5 per cent leaf powder concentration had caused maximum mortality (76%) of the weevil followed by 3 per cent with 62% and 5 percent with 48% mortalities respectively.

Keywords : Milk Weed Plant, Rice Weevil, Flower Powder.

1. INTRODUCTION

Milk weed plant (*Calotropis gigantea* R. Br), a notorious wasteland weed is known to be insecticidal (Solunke and Deshpande, 1991) and antifeedant (Pari *et al.,* 1998). The plant is found to possess different types of cardenolides that contain the arrow poisons such as calotropin (II), caloctin (IV) and calotoxin (V) (Roeske *et al.,* 1976). In India, most of the food grains produced are being stored at farmers premises under the most primitive conditions of storage and thus they are easily accessible to attack by a variety of insects pests and other agents in storage (Radhakrishnan *et al.,* 1983) The rice weevil, *Sitophilus oryzae* L.is serious insect pest of paddy grains under storage and is largely responsible for damage and frequently harbouring in stores, mills and warehouses. In the above background an attempt was made to find whether the *calotropis* plant is effective against rice weevil in storage.

2. MATERIALS AND METHODS

A laboratory study was conducted at the Department of Entomology, Faculty of Agriculture, Annamalai University with seven treatments of *Calotropis* leaf and flower powders and evaluated against therice weevil. The treatment consisted of leaf powder and flower powder of calotropis at 1%, 3% and 5% dosages in each. Observations on the number of adult weevils were made. The initial count was taken a day prior to the treatment and the post treatment counts were made at 1st, 3rd and 5th day after dusting.

3. RESULTS AND DISCUSSION

Among the plant parts of *C. gigantea* tested, the leaf powder at 5 per cent gave maximum mortality of *S. oryzae* (76 %) followed by 3 per cent leaf powder and flower powder 5 per cent effected 62% and 48% reduction respectively. Leaf powder 1% and flower powder 1% were found to be least effective against the rice weevil. (Table 1)

Table 1 : Efficacy of various parts of *C. gigantea* against rice weevil in stored rice

Treatments	Per cent Mortality at			
	1 DAT	3 DAT	5 DAT	Mean
Leaf powder 1%	30 (30.47)	34 (34.67)	38 (38.98)	34 (34.67)d
Leaf powder 3%	57 (60.67)	60 (64.35)	69 (76.15)	62 (66.87)b
Leaf powder 5%	68 (72.08)	74 (83.31)	86 (81.55)	76 (86.33)a
Flower powder 1%	2 (2.14)	6 (4.96)	10 (10.02)	6 (4.96)e
Flower powder 3%	16 (16.07)	20 (20.14)	24 (24.24)	20 (20.14)e
Flower powder 5%	44 (45.56)	48 (50.07)	52 (54.69)	48 (50.07)c
Untreated control	0 (0.00)	0 (0.00)	0 (0.00)	0 (0.00)

Values followed by diffierent letters are significant at $p < 0.05$ DMRT S.Ed. 4.03
DAT - Days after treatment CD 8.80

Parrish (1994) reported that by mixing leaf powder of guava (*Psidium guajava*) and eucalyptus (*Eucalyptus globulas*) reduced the infestation of *S. oryzae.*

References

Pari, K., P.J. Rao, C. Devakumar and J.N. Rastogi, 1998. A novel insect antifeedant non protein amino acid from *Calotropis gigantea. J. Natl. Prod.,* 61(1): 102-104.

Parrish, A., 1994. Indigenous post harvest knowledge in an egyptian oasis. *Indigenous knowledge and Development Monitor,* 2(1): 23.

Radhakrishnan, S., B. Rajamanickam and R.S. Madhav, 1983. Efficacy of certain insecticides against *Callosobruchus chinensis* (L.) on *Cajanus cajan.* Seeds in storage. *Ind. J. Ento.,* 34(1): 50-51.

Roske, C.N., J.N. Seibar, L.P. Brower and C.M. Moffit, 1976. Milkweed cardenolides and their comparative processing by monarch butterflies pp. 93-164. In: Wallace J.W. and Mansell, R.R. (Ed.) Recent advances in phytochemistry, biochemical interaction between plants and insects Plenum Press, New York.

Solunke, B.R. and S.V. Deshpande, 1991. Studies on use of plant products for control of lemon butterfly larvae. *J. Maharashtra Agri. Univ.,* 16(2): 302-303.

— xxx —

19

Evaluation of Certain Botanicals as Grain Protectants Against Rice Weevil, *Sitophilus oryzae* L.

JAYAKUMAR, M. AND S. JOHN WILLIAM
School of Entomology and Centre for Natural Resources Management, P.G. & Research Department of Advanced Zoology Biotechnology, Loyola College, Chennai- 600 034.

SUMMARY

Evaluation of crude extracts of four plant species viz. *Artemisia vulgaris, Argemone mexicana, Sphaeranthus indicus* and *Tephrosia purpurea* were obtained by using hexane, chloroform and ethyl acetate. Evaluation of various solvent extracts of these plants clearly indicated their varying degree of efficacy against *S. oryzae*. Significant ovicidal activity was observed in hexane extract of *S. indicus* and ethyl acetate extract of *T. purpurea* (63.33%). Maximum mortality of the weevil was observed in hexane extracts of *T. purpurea* and *A. mexicana*. Significant repellent activity was observed in ethyl acetate extracts of *S. indicus* followed by *A. vulgaris*.

Key words: Antifeednt, Egg Hatchability, Ovicidal Activity.

1. INTRODUCTION

Botanical pesticides have the advantage of providing novel modes of action against insects that can reduce the risk of cross-resistance as well as offer new leads for design of target specific molecules. In recent years, large numbers of plant extracts and secondary metabolites have been screened for their activities against stored product insects and they are reported to possess insecticidal, repellent or antifeedant properties against stored product insects (Grainage and Ahmed, 1988; Arnason *et al.,* 1989; Weaver and Subramanyam, 2000; Khatune *et al.,* 2002; Jayakumar *et al.,* 2005; Prabuseenivasan *et al.,* 2006). Although a large number of plant products in various forms have been screened against major pests of stored grain, there are many more species remain unexplored (Tiwari, 1993). Therefore, in the present study it was aimed to study the efficacy of four indigenous plant spescies viz, against *Sitophilus oryzae* (L.) a key pest of rice seeds in storage.

2. MATERIALS AND METHODS

2.1. Insect culture establishment

S.oryzae was collected from the local market and a culture was established in plastic containers (11.5 X 6 cm) containing whole soft wheat. The container was covered with muslin cloth and culture was maintained at 27± 2° C, 65-70% RH and light: dark cycle of 8:16 h. All the experiments were carried out under the same environmental conditions. Adult insects of 1-2 week after eclosion, were used.

2.2. Preparation of solvent extracts

Fresh leaves of the selected plants were collected in and around Chennai, Tamil Nadu, shade dried at room temperature (28± 2°C) and powdered using electric blender. The powder material (200 g) was sequentially extracted with 1000 ml of solvents such as hexane, chloroform and ethyl acetate and kept in shaker for 12h. After thorough mixing, the extract was filtered and the solvent was allowed evaporated using rotary vacuum evaporator to obtain a final solvent free residue. Appropriate concentrations (0.5%, 1%, 2.5% and 5%) were prepared by mixing the solvent residue with acetone and used in the experiments.

2.3. Insecticidal activity

Film residue method: A Film of extracts was prepared on glass petridishes (7cm dia.) by spraying under Potter's tower with one ml of each concentration (0.5, 1, 2.5 & 5%) at a pressure of 340-g/sq. cm and dried for about one hour. Twenty unsexed adults obtained from the culture were released in to each petridish. Mortality was recorded every 24 h for 4 days and the percentage of mortality was computed using the Abbott's formula. Untreated glass petridishes with insects used as control. Three replications were maintained in each.

2.4. Ovicidal activity

Ovicidal activity of extracts of the selected plant species against *S. oryzae* was studied with 100 ng freshly emerged adults in to batches of 100g wheat at 14% moisture content and incubated 30°C for 1 week. The adults then were removed and the grains were admixed with different concentrations of plant extracts. Then the wheat grains were dried and incubated for 7 weeks and number of young ones emerged out from each treatment was counted. Three replications were maintained for each concentration and untreated control.

2.5. Repellent activity

Glass olfactometer was used to find out the repellent activity of the plant extracts against *S. oryzae*. A multi-arm olfactometer consisting of middle glass chamber (5 cm diameter) from which project out wards five glass tubes (15cm length and 2 cm diameter) with equal gap 90° between them was used. The end of each arm was fitted with a plastic jar (8 X 6.5 cm). Filter paper strips (6X4 cm) were dipped in 5% concentration of each solvent extracts and air-dried and placed in the plastic jars of olfactometer and covered with muslin cloth. Filter paper dipped in acetone served as control. Fifty newly emerged adults of *S. oryzae* were introduced into the olfactometer through the central opening. After 1hour, the number of beetles found in each jars was recorded A and the percent repellency and the Excess Proportion Index (EPI) was calculated using the formula of Sakuma and Fukami (1985).

3. RESULTS

Laboratory evaluation of different solvent extracts of the selected plants indicated their varying degree of efficacy against *S. oryzae*. In general all the selected plant extracts showed some degree of activity. Results revealed that ethyl acetate extract of *T. purpurea* showed maximum ovicidal activity (66.33%), followed by hexane extract of *S. indicus* (63.33%) and *T. purpurea* (59.00%) at 5% concentration (Table 1). Notably all the three extracts of both *S. indicus* and *T. purpurea* showed more than 50% of ovicidal activity at higher concentration. *T. purpurea* accounting of 56 to 66.33% activity whereas in *S. indicus* it was ranging about 55 to 63.33% of ovicidal activity. *A. vulgaris* showed comparatively very less ovicidal activity against *S. oryzae.*

Table 1 : Ovicidal activity of different solvent based plant extracts against *Sitophilus oryzae* in wheat

Plant species	Solvents	% mortality at Concentrations			
		0.625%	1.25%	2.5%	5%
	Hexane (14.89)a	6.67 (18.43)a	10.00 (19.91)b	11.67 (27.69)b	21.67
Artemisia vulgaris	Chloroform (16.74)a	8.33 (19.91)a	11.67 (26.57)a	20.00 (37.46)a	37.00
	Ethyl acetate (14.89)a	6.67 (21.39)a	13.33 (27.69)a	21.67 (35.43)a	33.67
	Hexane (10.47)c	3.33 (21.39)c	13.33 (31.05)c	26.67 (35.67)b	34.00
Argemone mexicana	Chloroform (27.69)a	21.67 (36.27)a	35.00 (44.03)a	48.33 (47.64)a	54.67
	Ethyl acetate (24.12)b	16.67 (32.14)b	28.33 (39.23)b	40.00 (46.89)a	53.33
	Hexane (25.33)a	18.33 (35.24)a	33.33 (47.87)a	55.00 (52.71)a	63.33
Sphaeranthus indicus	Chloroform (21.39)b	13.33 (30.00)b	25.00 (40.16)b	41.67 (47.87)b	55.00
	Ethyl acetate (19.91)c	11.67 (25.33)c	18.33 (39.23)b	40.00 (48.45)b	56.00
	Hexane (19.91)b	11.67 (27.69)b	21.67 (41.15)b	43.33 (50.18)b	59.00
Tephrosia purpurea	Chloroform (19.91)b	11.67 (24.04)b	16.67 (32.14)c	28.33 (48.45)c	56.00
	Ethyl acetate (28.86)a	23.33 (37.23)a	36.67 (45.92)a	51.67 (54.51)a	66.33

Values in parentheses are angular transformed.
Values within a column different alphabets is statistically significant ($p<0.05$) by LSD.

Adult mortality was variable among the plants tested (Fig.1). Though there was no or low initial mortality up to 12h it gradually increased at the later durations. There was no mortality of the insect in the untreated control. In general, hexane and ethyl extracts showed higher activity at 96 h duration than the chloroform extracts. Significant ($P<0.05$) adult mortality was recorded in hexane extracts of *T. purpurea* and *A.mexicana* (74% and 70% respectively) followed by ethyl acetate extracts of *T. purpurea* (64%), *S. indicus* (54%) and *A. vulgaris* (54%).

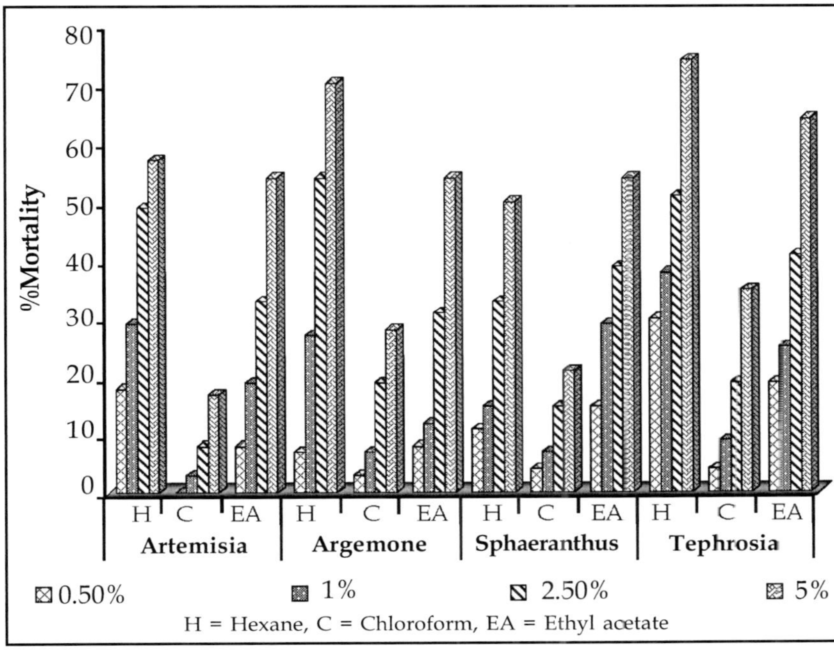

Fig 1 : Insecticidal activity of the selected plant species against *Sitophilus oryzae* at 96 hr

Figure 2 represents the repellent activity of the selected plant extracts against *S. oryzae*. Repellent activity of the selected plants was wide-ranging. In general, the ethyl acetate extract showed higher repellency than the other two solvent extracts. Percentage of repellent activity was ranged from 4.84% to 66.23%. Maximum repellent activity observed with the ethyl acetate extract of *S. indicus* (66.23%) and *A. vulgaris* (54.55%) followed by hexane extracts of *A. vulgaris* and *S. indicus* (42.65%) and the rest had poor repellent activity against *S. oryzae*.

4. DISCUSSION

The findings are suggestive of the adverse effects of crude solvents extracts of the selected plants against *S. oryzae*. The observed effects included egg hatchability, mortality and repellent activity. Present study revealed that ethyl acetate extract of *T. purpurea* showed maximum ovicidal activity followed by hexane extract of *S. indicus* and *T. prupurea*. This might be due to interference of the plant extracts with the eggs contour and cytoplasmic disintergration. Further, the egg mortality probably takes place due to blocking of oxygen supply which might inhibit further embryonic development. Similar physiological interference was also reported in *Azadirachta indica* (Mahgoub, 1992), *Acorus calamus* (Chiranjeevi and Sudhakar, 1996) and Prabuseenivasan *et al,* (2006).

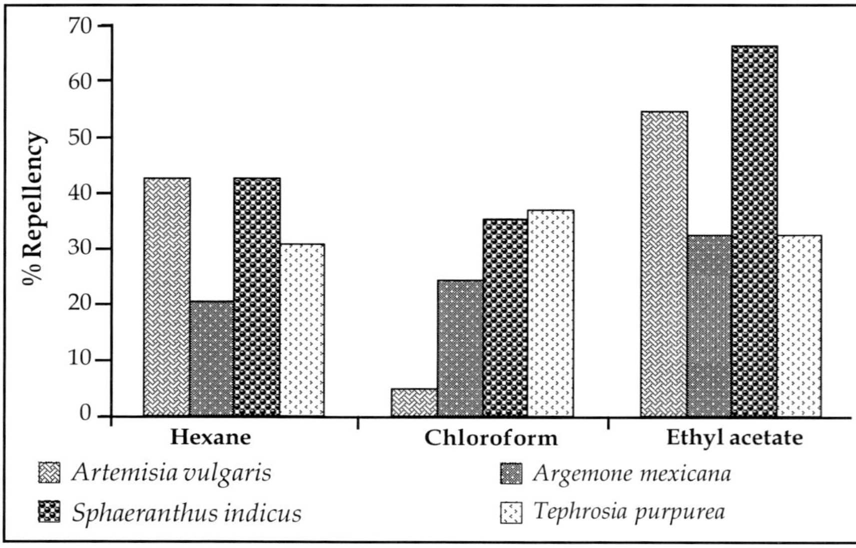

Fig. 2: Repellent activity of different solvent extracts of the selected plants against *S.oryzae*

Control of an adult insect is an important biological parameter to protect post harvest products. If the plant extracts have the insecticidal compounds that kill the adult insects, the egg-laying and progeny development are consequently reduced. Adulticidal activity of (E)-anethole, (+)-fenchone and estragole against *S. oryzae, Callosobruchus chinensis* L. and *Lasioderma serricorne* Fab. is dependent upon both dose and exposure time (Kim and Ahn, 2001). However, El-Nahal *et al.* (1989) stated that the exposure period appeared to be the more important factor to adults of five stored - product insect species than the dosage. In the present study the adult mortality was maximum at higher concentration and longer exposure period in hexane extracts of *T. purpurea* and *A. mexicana*.

Repellent activity increases the potential value of the materials in protecting grains from the attack by stored product insect (Bekele *et al.*, 1996). The toxic and repellent effects of phytochemicals on stored grain pests depend on several factors like chemical compositions of the plant products and insects susceptibility (Casida, 1990). There are a few techniques for determining if substances repel or attract stored-product insects. Use of filter paper or glass test arenas (Burkholder, 1990; Dowdy *et al.*, 1993) does not allow testing products on the grain. The present study showed strong repellent activity in the ethyl acetate extract of

T. purpurea followed by *A. vulgaris* against *S. oryzae*. Similar result was obtained by Prabuseenivasan *et al.* (2004) on the ethyl acetate extract of *Citrullus colocynthis.*

From this study, it is quite obvious that the bioactive principles present in *T. purpurea* and are responsible for egg hatchability, toxicity and repellent activity towards *S.indicus*. Hence it can be considered as one of the components in the IPM programmes, which could further reduce the application of the synthetic chemicals. However, further studies are needed to isolate the potential compound responsible for these biological activities.

Acknowledgement

The first author is thankful to the Department of Science and Technology, Govt. of India, New Delhi for providing financial support to carry out this work.

References

Arnason, J.T., B.J.R. Philoglene and P. Morand. 1989. Insecticides of Plant Origin. ACS Symposium Series No. 387. American Chemical Society, Washington, DC, 213 pp.

Bekele, A.J., D. Obeng-Ofori and A. Hassanali. 1996. Evaluation of *Ocimum kenyense* (Ayobangira) as source of repellents, toxicants and protectants in storage against three stored product insect pests. *J App. Ent.*, 121:169

Burkholder, W.E. 1990. Practical use of pheromones and other attractants for stored-product insects. In: Ridgway, R.L., Silverstein, R.M., Inscoe, M.N., (Eds.), Behavior-Modifying Chemicals for Insect Management: Applications of Pheromones and Other Attractants. Marcel Dekker, Inc., New York, pp. 497–516.

Casida, J.H. 1990. Pesticide mode of action, evidence for implications of a finite number of biochemical targets. In: Casida, J.E., (Ed.), Pesticides and Alternatives. Innovative Chemical and Biological Approaches to Pest Control, Elsevier, Amsterdam, pp. 11–22.

Chiranjeevi, C and T.R. Sudhakar. 1996. Effect of indigenous plant materials on the fecundity, adult emergence and development of pulse beetle, *Callosobruchus chinensis* (L.) in black gram. *Journal of Research-APAU*, 24(3&4), 57-61.

Dowdy, A.K., R.W. Howard, L.M. Seitz and W.H. McGaughey. 1993. Response of *Rhyzopertha dominica* (Coleoptera: Bostrichidae) to its aggregation pheromone and wheat volatiles. *Environ. Entomol.* 22: 965–970.

El-Nagal, A.K.M., G.H. Schmidt and E.M. Risha. 1989. Vapours of *Acorus calamus* oil: A space treatment for stored-product insects. *J. Stored Prod. Res.*, 25: 211-216.

Grainage, M and S. Ahmed. 1988. Handbook of plants with pest-control properties. Wiley, New Yark.

Jayakumar, M., K. Elumalai, A. Jeyasankar, N. Raja and S. Ignacimuthu, 2005. Biological activity of *Hyptis suaveolens* Poit (Lamiaceae) and *Melochia chorcorifolia* L. (Sterculiaceae) on cowpea weevil, *Callosobruchus maculatus* (F.) (Coleoptera: Bruchidae). *J. Ent. Res.*, 29(4): 265-269.

Khatune, N.A., M.E. Islam, M.A.A. Rahman, M.A. Baki, G. Sadik and M.E. Haque. 2002. Pesticidalactivity of a Novel coumestan derivative isolated from *Psoralea corylifolia* Linn. Against *Tribolium castaneum* Herbst. Adults and larvae (Coleoptera: Tenebrionidae). *Pakistan Journal of Agronomy*, 1(4): 112-115.

Kim, D.H and Y.J. Ahn. 2001. Contact and fumigant activities of constituents of *Foeniculum vulgare* fruit against three coleopteran stored-product insects. *Pest Management Science*, 57: 301–306.

Mahgoub, S. M. 1992. Neem seed extracts and powders as grain protectants to cowpea seeds against the cowpea weevil *Callosobruchus maculatus* Fab. Egyptian *J. Agri. Res.*, 70: 487-496.

Prabuseenivasan, S., M. Jayakumar, N. Raja and S. Ignacimuthu, 2004. Efficacy of bitter apple (*Citrullus colocynthis*) seed extracts against pulse beetle, *Callosobruchus maculatus* Fab. (Coleoptera: Bruchidae). *Entomon*, 29(1): 81-84.

Prabuseenivasan, S., M. Jayakumar, N. Raja and S. Ignacimuthu, 2006. Evaluation of bitter apple (*Citrullus colocynthis*) fruit extracts against pulse beetle, *Callosobruchus maculatus* Fab. (Coleoptera: Bruchidae), pp63-69. In: Green Pesticides for Insect Pest Management. Narosa Publishing House, New Delhi.

Tiwari, S.N. 1993. Efficacy of some plant products as grain protectant against *Sitophilus oryzae* (L.). *J. Insect Sci.*, 6(1): 158-160.

Weaver, D.K and B. Subramanyam. 2000, Botanicals. pp.303-320 In: Subramanyam, Bh., Hagstrum, D.W. (Eds.), Alternative to Pesticides in Stored-products IPM. Kluwer Academic Publishers, Massachusetts, USA.

— xxx —

20

Effect of Some Traditional Oils as Ovipositional Deterrents Against Pulse Beetle, *Callosobruchus maculatus* (Fab.) (Bruchidae: Coleoptera) on Green Gram

SATHYASEELAN, V.[1], V. BASKARAN, K. NATARAJAN AND S. MOHAN
Department of Agricultural Entomology, Tamil Nadu Agricultural University, Coimbatore-3, [1]Lecturer, Department of Entomology, Faculty of Agriculture, Annamalai University, Annamalai Nagar- 608002

SUMMARY

Results on the evaluation of some vegetable oils against pulse beetle revealed that both the neem oil (90.0%) and palmarosa oil (90.0%) @ 10 ml. /kg. of seed recorded highest ovipositional deterrency for three months followed by pongam oil (84.3%), madhuca oil (84.0%), castor oil (81.3%) and patchouli oil (79.3%). The least detterency was observed in the case of mustard oil (75.9%) and rice bran oil (72.4%). In general, traditional oils @ 5 ml. /kg. of seed showed a moderate ovipositional deterrency against the bruchid beetles for about 2 months after treatment whereas, oils @ 2.5 ml./kg. of seeds recorded least ovipositional deterrent effect on the pulse beetle in green gram.

Key words: Traditional Oils, Ovipositional Deterrency, Age-old Practices.

1. INTRODUCTION

Green gram, *Vigna radiata* (L.) is one of the important pulse crops mainly used for culinary and nowadays in cosmetic purposes. After harvest, green gram is stored as dry seeds for short to longer durations for use as seed material for the forthcoming season or for simply waiting of enhanced market price. The pulse beetle, *Callosobruchus maculatus* (Fab.) is the most dominant and destructive pest of stored green gram in India, causing 10.67 to 94.66 percent grain infestation. (Bhargava and Meena, 2002, Sonata *et al.*,2006 and Kumar *et al.*,2007). The beetle is not only effecting quantitative but also qualitative losses which degrade in its quality and makes it unfit for marketing. Treatment of the pulse seeds with vegetable oils prior to storage to protect them from stored product pests damage is an age-old practice in India and in African tribals too (Tapondjou *et al.*, 2002, Kim *et al.*,2003, Koona and Dorn,2005).Therefore the present investigation was carried out to study the effect of different vegetable oils (viz., neem, madhuva, pungam, mustard, patchouli, palmarosa, castor and rice bran oil) apainst the pulse beetle *C. maculatus* on the of green gram seeds in storage under laboratory condition.

2. MATERIALS AND METHODS

The present study was conducted at the Department of Agricultural Entomology, Tamil Nadu Agricultural University, Coimbatore, Tamil Nadu, during the year 2003 - 2004. The seeds of green gram were treated with some traditional oils *viz.*, neem oil (*Azadirachta indica* An.dr.Juss.), madhuca oil (*Madhuca latifolia* J.Gmelin.), pongam oil (*Pongamia glabra* L.), mustard oil (*Brassica juncea* (L.) Czernj.), patchouli oil (*Pogostemon patchouli*), palmarosa oil (*Cymbopogon martini* var. *motia*), castor oil (*Ricinus communis* L.) and rice bran oil (*Oryza sativa* L.) to determine their ovipositional deterrent effect at different doses viz., 2.5 ml, 5.0 ml and 10.0 ml/ kg of seeds against pulse beetle *C. maculatus*. Three samples of 25g seeds each were drawn from each treatment and kept in plastic containers. Two pairs of one to two days old adults were released in to each plastic container of 7.00 x 5.00 cms size and was covered with a perforated lid. In the same manner an untreated control was also maintained separately as indicated by Kayode (2006), Abeywickrama *et al.* (2006) and Negahban *et al.* (2007).

The experiment was conducted at 30± 1°C and 70 ± 5 per cent relative humidity under laboratory condition in a Completely

Randomized Design with three replications. Observations were made on the ovipositional deterrency against the pulse beetle.

3. RESULTS AND DISCUSSION

The results revealed that the treatments of neem oil and palmarosa oil (90.0%) recorded highest ovipositional deterrency irrespective of the dose, whereas pongam oil (84.3%), madhuca oil (84.0%) and castor oil (81.3%) stood followed by patchouli oil (79.3%) which showed moderate ovipositional deterrency (Table 1). Least ovipositional deterrency was noticed in the case of mustard oil (75.9%) and rice bran oil (72.4%) against the pulse beetle.

Table 1 : Ovipositional deterrency of certain traditional oils against pulse beetle in green gram

Treatments	Dose ml / kg of seeds	Percent ovipositional deterrency at		
		1 month	2 months	3 months
Neem oil	2.5	90.0^{a} (100.0)	75.8^{bc} (94.0)	65.2^{e} (82.3)
	5.0	90.0^{a} (100.0)	75.8^{bc} (94.0)	66.3^{e}(83.0)
	10.0	90.0^{a} (100.0)	90.0^{a} (100.0)	90.0^{a} (100.0)
Madhuca oil	2.5	59.2^{g} (73.8)	56.7^{fg} (69.9)	33.4^{m} (30.2)
	5.0	71.6^{e} (90.0)	67.5^{d} (85.4)	63.2^{ef} (79.7)
	10.0	84.5^{ab} (99.0)	84.3^{ab} 99.0)	84.0^{ab} (99.0)
Pongam oil	2.5	61.3^{fg} (76.9)	50.6^{hi} (59.7)	44.1^{k} (48.3)
	5.0	62.1^{fg} (78.0)	52.5^{hi} (63.0)	53.6^{hi} (64.8)
	10.0	90.0^{a} (100.0)	86.2^{ab} (99.0)	84.3^{ab} (99.0)
Mustard oil	2.5	58.5^{g} (72.3)	48.8^{ij} (56.5)	44.5^{kl} (48.5)
	5.0	71.6^{e} (90.0)	61.3^{fg} (76.9)	53.2^{jk} (64.0)
	10.0	85.2^{ab} (99.0)	78.0^{b} (95.6)	75.9^{cd} (94.1)
Patchouli oil	2.5	50.2^{jk} (60.0)	57.3^{ef} (70.9)	55.4^{ij} (67.7)
	5.0	74.8^{de} (93.2)	66.8^{c} (84.5)	50.8^{jk} (60.0)
	10.0	90.0^{a} (100.0)	74.8^{de} (93.2)	79.3^{bc} (96.6)
Palmarosa oil	2.5	64.7^{f} (81.7)	48.8^{i} (56.5)	34.7^{m} (32.5)
	5.0	79.5^{bc} (96.7)	65.0^{d} (82.1)	57.4^{fg} (71.0)
	10.0	90.0^{a} (100.0)	90.0^{a} (100.0)	90.0^{a} (100.0)
Castor oil	2.5	53.7^{gh} (65.0)	50.1^{j} (59.5)	50.1^{ij} (59.5)
	5.0	76.1^{cd} (94.2)	60.4^{e} (75.6)	56.6^{hi} (69.7)
	10.0	90.0^{a} (100.0)	85.2^{ab} (99.0)	81.3^{bc} (97.7)
Rice bran oil	2.5	39.0^{h} (39.6)	26.1^{l} ((19.3)	24.3^{k} (16.2)
	5.0	59.5^{g} (74.2)	53.7^{gh} (65.0)	51.1^{jk} (50.6)
	10.0	73.9^{c} (92.3)	71.6^{e} (90.0)	72.4^{e} (90.8)
Control		0.0	0.0	0.0
SEm(±)		1.3	1.3	1.9
CD at 5%		3.8	3.8	5.3
CV %		2.9	3.5	5.3

Rice bran oil (39.6%) @ 2.5 ml/ kg seed showed less deterrency effect when compared to the remaining traditional oils at the same dosage. Even at lower dose level, neem oil exerted highest ovipositional deterrency (90.0%) followed by palmorosa oil (64.7%), pongam oil (61.3%) and madhuca oil (59.2%) was on par with mustard oil (58.5%). Castor oil (53.7%) and patchouli oil (50.2%) ranked next and the least deterrency was recorded with rice bran oil @ 2.5 ml/ kg seed during one month after treatment against pulse beetle. Similar results were recorded in different traditional extracts and oils by Keita *et al.,* (2001), Adedire and Akinneye, (2004), Cherry *et al.,* (2007) and Koona *et al* ., (2007).

Among the oils, neem oil @ 5ml/ kg of seed, showed the highest deterrency (90.0%) followed by palmarosa oil (79.5%). Castor oil and patchouli oil were on par with each other(76.1%, 74.8%), whereas mathuca oil and mustard oil (71.6%) showed same level of deterrency against the pulse beetle. The least deterrency was recorded with pongam oil (62.1%) followed by rice bran oil (59.5%) at the same dose level.

At a higher dose of 10 ml/ kg of seed, neem oil, palmarosa oil, castor oil, patchouli oil and pungam oil gave total protection of seeds recording cent percent deterrency against the pulse beetle. Mustard oil and madhuca oil behaved alike. Seeds treated with neem oil and palmarosa oil @ 10ml / kg seed maintained their superiority over rest of the treatments even after two months of storage and showed highest deterrency (90.0%) and in turn gave complete protection against the pulse beetle. Thus pongam oil, castor oil and madhuca oil (84.3%) ranked second in deterrency and offered good protection followed by patchouli oil and mustard oil (78.0%).

After two months of storage, all the oils @ 2.5 and 5.0 ml / kg seed except neem oil (all the three doses) were found less effective recording 48.8 to 57.3 per cent deterrency whereas rice bran oil recorded very less efficacy about 26.1 to 53.7 per cent respectively for 2.5 and 5.0 ml / kg seed against pulse beetle. The results of the present study confirm findings of Babu *et al.* (1999). Boeke *et al.* (2004) and Kestenholz *et al.* (2007) in various plant extracts and oils.

Most of the traditional oils lost their efficacy on prolonged storage. However, neem oil and palmarosa oil at the highest dose maintained their effectiveness for three months recording maximum deterrency with almost complete protection against the pulse beetle. Pongam oil, madhuca oil and castor oil @ 10.0 ml / kg seed showed deterrency with a protection of 84.3 to 81.3 per cent for three months Our finding is in agreement with Raghvani and Kapadia (2003) on pigeon pea and Keita *et al.* (2000), Ketoh *et al.* (2005), Rahman and Talukder (2006) and Upadhyay *et al.* (2006) in other pulses.

Therefore, the results revealed that mustard oil at initial stage under all the doses levels showed considerable protection against the pulse beetle but as the duration increased, its deterrent effect decreased.

Neem oil and palmarosa oil @ 10.0 ml / kg seed recorded maximum deterrency even after three months. Castor oil, madhuca oil and pungam oil showed moderate deterrency effect against pulse beetle upto three months of storage.

References

Abeywickrama, K.Adhikari, A.A.C.K.Paranagama, P.and Gamage, C.S.P. 2006.The efficacy of essential oil of *Alpinia calcarata* (Rosc.) and its major constituent, 1,8-cineole, as protectants of cowpea against *Callosobruchus maculatus* (F.) (Coleoptera: Bruchidae). *Can. J. Plant Sci.* 86(3): 1-827.

Adedire, C.O. Akinneye, J.O. 2004. Biological activity of tree marigold, *Tithonia diversifolia*, on cowpea seed bruchid, *Callosobruchus maculatus* (Coleoptera: Bruchidae). *Ann. Appl. Biol.*, (144)2: 185-189.

Babu, A. Raja, N. Albert, S. Ignacimuthu, S. Dorn, S. 1999. Comparative efficacy of some indigenous plant extracts against the pulse beetle *Callosobruchus maculatus* F. (Coleoptera: Bruchidae). *Biol. Agri. Horti.*, (17)2: 145-150.

Boeke, S.J. Kossou, D.K. Huis, A. van Loon, J.J.A. van Dicke, M. 2004. Field trials with plant products to protect stored cowpea against insect damage. *Inter. J. Pest Mgt.*, 50(1): 1-9.

Cherry, A.J. Abalo, P. Hell, K. Korie, S. 2007. Farm-scale trials to compare the entomopathogenic fungus *Beauveria bassiana* with pirimiphos methyl + deltamethrin and essential oil of lemon grass for protection of stored cowpea against *Callosobruchus maculatus* (Coleoptera: Bruchidae). *Tests of Agrochemicals and Cultivars*, 151(1): 1-1

Kayode, J.2006. Pesticidal activity of the leaf and bark extracts of *Gliricidia sepium* on *Callosobruchus maculatus*. *Journal of Sustainable Forestry*, 22(3-4): 57-61.

Keita, S.M. Vincent, C. Schmit, J.P. Ramaswamy, S. Belanger, A. 2000. Effect of various essential oils on *Callosobruchus maculatus* (F.) (Coleoptera: Bruchidae). *J. Stored Prod. Res.*, 36(4): 355-364.

Kestenholz, C. Stevenson, P.C. Belmain, S.R. 2007. Comparative study of field and laboratory evaluations of the ethnobotanical, *Cassia sophera* L.

(Leguminosae) for bioactivity against the storage pests, *Callosobruchus maculatus* (F.) (Coleoptera: Bruchidae) and *Sitophilus oryzae* (L.) (Coleoptera: Curculionidae). *J. Stored Prod. Res.*, 43 (1): 79-86.

Ketoh, G.K. Koumaglo, H.K. Glitho, I.A. 2005. Inhibition of *Callosobruchus maculatus* (F.) (Coleoptera: Bruchidae) development with essential oil extracted from *Cymbopogon schoenanthus* L. Spreng. (Poaceae), and the wasp, *Dinarmus basalis* (Rondani) (Hymenoptera: Pteromalidae). *J. Stored Prod. Res.*, 41, (4): 363-371.

Kim, S.I. Roh, J.Y. Kim, D.H. Lee, H.S. Ahn, Y.J. 2003. Insecticidal activities of aromatic plant extracts and essential oils against *Sitophilus oryzae* and *Callosobruchus chinensis*. *J. Stored Prod. Res*, 39 (3): 293-303.

Koona, P. and Dorn, S. 2005. Extracts from *Tephrosia vogelii* for the protection of stored legume seeds against damage by three bruchid species. *Ann. Appl. Biol.*, 147, (1): 43-48.

Koona,p. Tatchago,v. and Malaa,D. 2007. Impregnated bags for safer storage of legume grains in West and Central Africa : *J. Stored Prod. Res.* 43, (3): 248-251.

Kumar, R. Srivastava, M. Dubey, N.K. 2007. Evaluation of *Cymbopogon martinii* Oil Extract for Control of Postharvest Insect Deterioration in Cereals and Legumes. *J. Food Prot.*, 70, (1):172-178.

Negahban, M. Moharramipour, S.and Sefidkon, F.2007.Fumigant toxicity of essential oil from *Artemisia sieberi* Besser against three stored-product insects. *J. Stored Prod. Res.*, 43, (2): 123-128.

Raghvani,B.R. and Kapadia,M.N. 2003. Efficacy of different vegetable oils as seed protectants of pigeon pea against *Callosobruchus maculatus* (Fab.). *Ind. J. Plant Prot.*, 31, (1): 115-118.

Rahman, A. and Talukder, F.A.2006. Bioefficacy of some plant derivatives that protect grain against the pulse beetle, *Callosobruchus maculatus. J. Insect Sci.*, 6.

Sonata, P. Talekar, N.S.and Srinivas, P.2006. Characterization of *Callosobruchus chinensis* (L.) resistance in *Vigna umbellata* (Thunb.) Ohwi & Ohashi. *J. Stored Prod.* Res., 42, (3): 313-327.

Tapondjou, L.A. Adler, C. Bouda, H. Fontem, D.A. 2002. Efficacy of powder and essential oil from *Chenopodium ambrosioides* leaves as post-harvest grain protectants against six-stored product beetles. *J. Stored Prod. Res.* 38, (4): 395-402.

Upadhyay, R.K. Rohatgi, L.Chaubey, M.K. and Jain, S.C. 2006. Ovipositional responses of the pulse beetle, *Bruchus chinensis* (Coleoptera: Bruchidae) to extracts and compounds of *Capparis deciduas. J. Agri. Food Che.*. 54(26): 9747-9753.

— xxx —

21

Effect of *Acorus calamus* L. Rhizome Powder and Its Dust Formulations on Pulse Beetle, *Callosobruchus maculatus* (F.) (Coleoptera: Bruchidae)

GOVINDAN, K.[1] S. JEYARAJAN NELSON[2] AND P.M.M.DAVID[2]

[1]Ph.D., Scholar, Department of Agricultural Entomology, Tamil Nadu Agricultural University, Coimbatore-641 003

[2] Professors, Department of Agricultural Entomology, Agricultural College and Research Institute, Killikulam - 628 252

SUMMARY

Results showed that 100 per cent mortality was recorded in *A. calamus* (L.) rhizome powder 0.75, 1.00, 1.25 and 2.00 per cent at 48 hr after the treatment. A. *calamus* 10D (talc as filler) (Ac 10DT) caused 100 per cent mortality at 3.50 per cent after 72 hr of the treatment. *A. calamus* 10D (fly ash as filler) (Ac 10DF) caused 100 per cent mortality after 84hr of the treatment. *A. calamus* 2.00 per cent rhizome powder acted as the best ovipositional deterrent wherein only 7.00 eggs were laid compared to 265.00 eggs in control. No seed weight loss was observed. A. *calamus* 10D (Ac 10DT) and *A. calamus* 10D (Ac 10DF). Ac 10DT 4.00 per cent showed very good ovipositional deterrent effect with only 16.00 eggs compared to 204.00 eggs in untreated control. Few adults (7.66Nos) were emerged from the Ac 10DT 4.00 per cent. Seed weight loss was only 0.16 per cent in Ac 10DT 4.00 per cent compared to 18.38 per cent in untreated control. In Ac 10DF

4.00 per cent, only 18.00 eggs were laid while in untreated control 188.00 eggs were laid. Adult emergence was less (6.66) in Ac 10DF 4.00 per cent. Seed weight loss was minimum (0.11%) in Ac 10DF 4.00 per cent. Germination was maximum (89.33%) in Ac 10DF 4.00 per cent.

Key words : Acorus calamus, Rhizome Powder, Dust Formulations.

1. INTRODUCTION

Bruchids are known to inflict quantitative and qualitative losses to stored pulses (Adams and Schulten, 1978). Use of synthetic pesticides especially insecticides since last four decades led to wide spectrum of pest problems like pest resistance to pesticides, resurgence of pests, residues in food and soil and risks to human and animal health besides environmental pollution (Mohapatra and Gupta, 1998). Ketkar (1987) reported the insecticidal and grain protectant properties of different plants. Many plants were reported to possess activities against stored grain pests. Hence dust formulations of *Acorus calamus* L. rhizome powder were evaluated for oviposition, adult emergence, seed damage by *Callosobruchus maculatus* F. and seed germination.

2. MATERIALS AND METHODS

Laboratory experiments were carried out at Agricultural College and Research Institute, Killikulam during 2003 – 2005. *C. maculatus* was reared on black gram seeds in glass jars covered with muslin cloth by following the method developed by Credland and Wright (1989). *A. calamus* rhizomes were procured, shade dried, powdered using a pulverizer and used in the study. Twenty grams of black gram seeds at 12 per cent moisture content were taken in petridishes. The *A. calamus* rhizome powder and 10 % dust formulation (w/w) were added to black gram seeds and shaken thoroughly. Thirty newly emerged adults were released into each petridish and kept in the laboratory. Mortality was recorded at 24 hr intervals in seven days. Three replications were maintained in each treatment. *A. calamus* rhizome powders was made as 10 per cent dust formulation using different fillers *viz.*, talc and fly ash. The effect of the new dust formulations *viz., A. calamus* 10 D (talc as filler) (Ac 10DT) and *A. calamus* 10 D (fly ash as filler) (Ac 10DF) were tested against *C. maculatus* at different concentrations *viz.*, 0.25, 0.50, 0.75, 1.00, 1.25, 1.50, 1.75 and 2.00 per cent as above.

Oviposition, adult emergence, seed damage and seed germination were observed. Twenty grams of black gram seeds at 12 per cent moisture content was taken in glass bottles. *A. calamus* rhizome powder and 10 % dust formulation (w/w) was added to black gram seeds and shaken thoroughly. Activated clay (1:100) was used as a check. Then the glass bottles were covered firmly using muslin cloth. Five pairs of newly emerged adults of *C. maculatus* were released in to each glass bottle, covered firmly and tested at ambient conditions. Three replications were maintained in each treatment. Number of eggs laid on the seeds was counted on third day after the release of beetles. On 5th day after the release of beetles counts on translucent unhatched eggs and opaque hatched eggs were taken and hatchability percentage was worked out. On 15 th day after the release of beetles all the dead insects were removed from the bottles to prevent them from getting mixed with the first generation (F1) offsprings. The number of newly emerged adults was counted and removed from the bottle once in three days till the complete emergence of F1 offspring (up to 60 days after treatment). Observations on number of damaged and undamaged seeds were taken on 45th day after the treatment and seed damage percentage was worked out. The weight of the seed was taken on 45th day after the treatment and seed weight loss percentage was worked out. After the experiment, germination of black gram seeds was assessed on 90th day after the treatment using roll towel method.

3. RESULTS AND DISCUSSION

Data on mortality of *C. maculatus* after 24, 48 and 72 hr of the treatment were recorded. Results revealed that *A. calamus* at 1.75 and 2.00 per cent was significantly different from all other treatments with 55.55 per cent mortality after 24hr of treatment (Table 1). All the beetles were dead in *A. calamus* 0.75, 1.00 1.25, 1.50, 1.75 and 2.00 per cent except in *A. calamus* at 0.25 and 0.50 per cent at 48 hr after the treatment. The results agree with the results of Khan and Borle (1985) who found that *A. calamus* rhizome could act as a contact or stomach poison, or antifeedant and repellent against the newly emerged adults. At 72 hr after the treatment all the doses of *A. calamus* 100 per cent mortality was observed.

A. calamus 10 D (talc as filler) (Ac10DT) was tested for their insecticidal action in comparison with activated clay 1.00 per cent and talc 4 per cent (Table 2). Mortality of *C. maculatus* after 12, 24, 36, 48, 60, 72, 84, 96, 108 and 120 hr after the treatment were recorded. Ac10DT 2.00 per cent excelled in its effect of with 49.99 per cent mortality followed by Ac10DT 1.50 per cent (44.44%) and Ac10DT 3.50 per cent (43.33%). The same trend was noticed with steady increase in the mortality up to 60hr after the treatment. All the beetles were dead in Ac10DT 3.50 per cent 72 hr after the treatment while in Ac10DT 4.00 per cent, 3.00 and 2.50 per cent 98.88, 94.44 and 92.21 per cent beetles were dead, respectively but, all the treatments were on par in their effect. At 84 hr after the treatment cent per cent mortality was seen in Ac10DT 4.00 per cent followed by Ac10DT 2.00 per cent (95.95%) and 1.50 per cent (92.21%). Cent per cent mortality was observed in Ac10DT 2.00 at 96 hr after the treatment. At 108 hr after the treatment 100 per cent mortality was observed in Ac10DT 1.00 per cent. At 120 hr after the treatment, 100 per cent mortality was observed in Ac10DT 0.50 per cent. Ac10 DT 4.00 and 3.5 per cent were significantly superior to all other treatments. Igantowicz and Wesolowska (1996) reported that *A. calamus* powder was repellent against stored product pests.

Table 1 : Insecticidal action of *Acorus calamus* rhizome powder on *Callosobruchus* maculatus

Treatments	**% Adult mortality at**			
	24 hr	**48 hr**	**72 hr**	**Mean**
A. calamus 0.25 %	19.77 (26.51)	82.21 (65.19)	100.00 (89.47)	67.33 (60.39)
A. calamus 0.50%	36.66 (37.25)	97.77 (82.80)	100.00 (89.47)	78.14 (69.84)
A. calamus 0.75%	39.99 (39.22)	100.00 (89.47)	100.00 (89.47)	79.99 (72.72)
A. calamus 1.00%	45.55 (42.44)	100.00 (89.47)	100.00 (89.47)	81.85 (73.80)
A. calamus 1.25%	46.66 (43.08)	100.00 (89.47)	100.00 (89.47)	82.22 (74.01)
A. calamus 1.50%	49.97 (44.99)	100.00 (89.47)	100.00 (89.47)	83.32 (74.65)
A. calamus 1.75%	55.55 (46.91)	100.00 (89.47)	100.00 (89.47)	85.18 (75.29)
A. calamus 2.00%	55.55 (48.18)	100.00 (89.47)	100.00 (89.47)	85.18 (75.71)
Control	4.44 (3.85)	8.88 (17.27)	17.77 (24.91)	10.36 (15.34)
Mean	39.35 (36.94)	87.65 (69.84)	90.86 (82.30)	72.62 (65.75)

Figures in parentheses are transformed arcsine values. Each value is a mean of 3 replications.

	SED	CD (0.05)
Treatment	0.96	1.94**
Period	0.55	1.12**
Treatment x period	0.67	3.36 *

Table 2 : Insecticidal action of *Acorus calamus* 10D (talc as filler) (Ac 10 DT) on *Callosobruchus maculatus*

Treatments	% Adult mortality at										Mean
	12 hr	24 hr	36 hr	48 hr	60 hr	72 hr	84 hr	96 hr	108 hr	120 hr	
Ac 10 DT 0.50 %	27.77 (31.74)	33.33 (35.24)	35.55 (36.59)	41.11 (39.87)	45.55 (42.44)	61.10 (51.44)	76.66 (61.11)	89.99 (71.72)	96.55 (81.20)	100.00 (89.47)	60.76 (54.08)
Ac 10 DT 1.00%	35.55 (36.58)	41.11 (39.87)	43.33 (41.16)	47.77 (43.72)	52.22 (46.27)	73.33 (58.93)	83.33 (65.97)	96.66 (81.32)	100.00 (89.47)	100.00 (89.47)	67.33 (59.28)
Ac 10 DT 1.50%	39.99 (39.22)	47.77 (43.72)	50.00 (45.00)	53.33 (46.91)	55.55 (48.18)	87.77 (69.57)	92.21 (74.36)	98.88 (86.14)	100.00 (89.47)	100.00 (89.47)	72.55 (63.20)
Ac 10 DT 2.00%	49.99 (44.99)	49.99 (44.99)	55.55 (48.18)	55.55 (48.18)	62.22 (52.07)	82.21 (65.19)	95.55 (79.85)	100.00 (89.47)	100.00 (89.47)	100.00 (89.47)	75.10 (65.19)
Ac 10 DT 2.50%	44.44 (41.80)	44.44 (41.80)	46.66 (43.08)	53.33 (46.91)	58.88 (50.12)	92.21 (74.36)	100.00 (89.47)	100.00 (89.47)	100.00 (89.47)	100.00 (89.47)	73.99 (65.60)
Ac 10 DT 3.00%	39.99 (39.22)	43.33 (41.16)	43.33 (41.16)	53.33 (46.91)	56.66 (48.83)	94.44 (76.51)	100.00 (89.47)	100.00 (89.47)	100.00 (89.47)	100.00 (89.47)	73.10 (65.17)
Ac 10 DT 3.50%	43.33 (41.16)	45.55 (42.44)	47.77 (43.72)	53.33 (46.91)	61.10 (51.43)	100.00 (89.47)	100.00 (89.47)	100.00 (89.47)	100.00 (89.47)	100.00 (89.47)	75.10 (67.30)
Ac 10 DT 4.00%	38.88 (38.57)	42.22 (40.52)	46.66 (43.08)	58.88 (50.12)	63.33 (52.74)	98.88 (86.14)	100.00 (89.47)	100.00 (89.47)	100.00 (89.47)	100.00 (89.47)	74.88 (66.91)
Activated clay1.00%	92.21 (89.47)	100.00 (89.47)	100.00 (89.47)	100.00 (89.47)	100.00 (89.47)	100.00 (89.47)	100.00 (89.47)	100.00 (89.47)	100.00 (89.47)	100.00 (89.47)	99.22 (87.96)
Talc 4.00%	3.33 (10.51)	5.55 (13.47)	8.88 (17.27)	14.44 (22.30)	15.53 (23.18)	28.88 (32.45)	37.77 (37.92)	41.11 (39.87)	52.22 (46.27)	56.66 (48.83)	
Control	4.44 (11.99)	6.66 (14.95)	11.10 (19.16)	14.44 (22.30)	18.88 (25.74)	25.55 (30.35)	27.77 (31.37)	43.33 (41.16)	45.55 (42.44)	48.88 (44.36)	24.66 (28.42)
Mean	41.66 (39.96)	45.44 (43.42)	47.99 (45.06)	53.10 (48.13)	57.44 (50.73)	81.55 (69.14)	87.55 (76.04)	92.88 (81.72)	94.21 (83.94)	94.88 (84.96)	

Each value is a mean of 3 replications. Values in parentheses are transformed arcsine values

	SED	CD (0.05)
Treatment	0.59	1.16 **
Period	0.59	1.16**
Treatment x period	1.86	3.68 **

Table 3 : Insecticidal action of *Acorus calamus* 10D (fly ash as filler) (A 10 DF) on *Callosobruchus maculatus*

Treatments	% Adult mortality at								Mean
	12 hr	24 hr	36 hr	48 hr	60 hr	72 hr	84 hr	96 hr	
Ac 10 DF 0.50%	11.11 (19.42)	22.22 (28.07)	35.55 (36.59)	50.00 (44.99)	56.66 (48.83)	67.77 (55.41)	89.99 (71.72)	100.00 (89.47)	54.16 (49.32)
Ac 10 DF 1.00%	12.22 (20.42)	24.44 (29.61)	36.66 (37.25)	50.00 (44.99)	65.55 (54.13)	75.55 (60.37)	95.55 (77.99)	100.00 (89.47)	57.49 (51.78)
Ac 10 DF 1.50%	15.55 (23.19)	27.77 (31.76)	38.88 (38.55)	47.77 (43.72)	61.10 (51.44)	75.55 (60.58)	97.77 (82.80)	100.00 (89.47)	58.05 (52.69)
Ac 10 DF 2.00%	15.55 (23.19)	29.99 (33.18)	41.10 (39.86)	56.66 (48.85)	71.10 (57.58)	93.33 (75.35)	100.00 (89.47)	100.00 (89.47)	63.46 (57.12)
Ac 10 DF 2.50%	21.11 (27.33)	29.99 (33.18)	42.22 (40.50)	63.33 (52.78)	71.10 (57.54)	89.99 (71.72)	100.00 (89.47)	100.00 (89.47)	64.71 (57.75)
Ac 10 DF 3.00%	23.32 (28.75)	31.10 (33.87)	44.44 (41.80)	58.88 (50.13)	75.55 (60.44)	91.10 (73.47)	100.00 (89.47)	100.00 (89.47)	65.55 (58.43)
Ac 10 DF 3.50%	19.99 (26.51)	28.88 (32.45)	50.00 (44.99)	73.33 (59.02)	80.00 (63.48)	93.33 (75.35)	100.00 (89.47)	100.00 (89.47)	68.19 (60.09)
Ac 10 DF 4.00%	23.33 (28.84)	29.99 (33.18)	56.66 (48.83)	81.10 (64.37)	90.00 (71.72)	97.77 (82.80)	100.00 (89.47)	100.00 (89.47)	72.35 (63.59)
Activated clay 1.00%	89.99 (71.72)	100.00 (89.47)	100.00 (89.47)	100.00 (89.47)	100.00 (89.47)	100.00 (89.47)	100.00 (89.47)	100.00 (89.47)	98.74 (87.26)
Fly ash 4.00%	4.44 (11.99)	11.11 (19.42)	21.10 (27.28)	38.88 (38.56)	47.77 (43.72)	61.10 (51.44)	67.77 (55.41)	75.55 (60.37)	40.96 (38.52)
Control	1.11 (3.85)	5.55 (13.47)	8.88 (17.27)	10.00 (18.43)	13.33 (21.41)	14.44 (22.30)	20.00 (26.56)	32.22 (34.57)	13.33 (19.73)
Mean	23.33 (27.32)	32.99 (35.83)	45.44 (43.51)	59.10 (51.68)	68.44 (57.61)	79.88 (66.68)	90.33 (79.59)	93.22 (83.98)	

Each value is a mean of 3 replications values in parentheses are transformed arcsine values

	SED	CD (0.05)
Treatment	0.77	1.52 **
Period	0.69	1.36**
Treatment x period	2.18	4.32 **

A. calamus 10 D with fly ash as filler (Ac10DF) at same doses as (Ac10DT) as tested for its insecticidal action of (Table 3) against *C. maculatus* after 12, 24, 36, 48, 60, 72, 84 and 96 after the treatment. Ac10DF 4.00, 3.00 and 2.50 per cent surpassed other treatments except activated clay with 23.33, 23.32 and 21.11 per cent mortality at 12 hr after the treatment, respectively and these treatments were significantly different from each other. At 24 hr after the treatment 100 per cent mortality was seen in activated clay 1.00 per cent which was followed by Ac10DF 3.00 per cent (31.10%), Ac10DF 2.00, 2.50 and 4.00 per cent (29.99%). At 72 hr after the treatment highest mortality was noted in Ac10DF 4.00 per cent (97.77%) followed by Ac10DF 2.00 and Ac10DF 3.50 per cent (93.33%). At 84 hr after the treatment 100 per cent mortality was registered in Ac10DF 2.00, 2.50, 3.00, 3.50 and 4.00 where as minimum mortality was observed in Ac10DF 0.50 per cent (89.99%). The present findings corroborates with those of Senthil kumar *et al.* (1998) fly ash has been evaluated for controlling pre harvest pest in soybean. All the beetles were dead in Ac10DF 0.50, 1.00 and 1.50 per cent at 96 hr after the treatment. In untreated control and fly ash 4 per cent, 32.22 and 75.55 per cent mortality was recorded at 96 hr after the treatment, respectively. Though ash (Golob and Webley, 1980; Golob, 1984; Stoll, 1988) was usually used in stored grain for protection against insect pests, Ac10DF 4.00 per cent was significantly superior than all other Ac10DF treatments but significantly lower than activated clay.

Though dust formulation of was tested against rice bug (Jeyarajan Nelson, et al., 2003) the effect of dust formulation against *C. maculatus* was not tested. Since, Ac10DT and Ac10DF 4.00 is equivalent to 0.40 per cent of rhizome powder, the insecticidal action was expressed. Hence, the findings on *A. calamus* rhizome powder could be taken as a support for the present finding. Shivanna *et al.* (1994) found that sweet flag rhizome powder mixed with red gram at 0.5, 1.5, 2.5g per 50g gave maximum protection against three generations of *C. chinensis.* Chiranjeevi (1991) protected green gram seeds from *C. chinensis* with *A. calamus* powder at 1.5 parts per 100 parts of seed. Pandey *et al.* (1976) reported that 100 parts of *Cicer arietinum* mixed with 2 parts powdered of sweet flag gave good protection against *C. chinensis* attack.

The pulse beetle laid significantly variable number of eggs on black gram seeds treated with different concentrations of *A. calamus* viz., 0.25, 0.50, 0.75, 1.00, 1.25, 1.50, 1.75 and 2.00 per cent (Table 4). Few eggs were laid by female in *A. calamus* 2.00 per cent (7.00Nos) followed by *A. calamus* 1.25 per cent (9.33Nos) and both the above treatments were at par in their effect. Among the different concentrations of *A. calamus,* maximum eggs were laid by female in *A. calamus* 0. 50 per cent (22.33 Nos). Rao *et al.* (1993) also reported that *A. calamus* powder had residual effects up to 90 days inhibiting oviposition and new progeny development. Hatchability was significantly less in seed treated with *A. calamus* 1.25 per cent. No adult beetle emergence was observed in all the treatments of *A. calamus* except in *A. calamus* 0.25 per cent. Regarding the seed weight loss no weight loss was observed in seed treated with *A. calamus* 0.50, 0.75, 1.00, 1.25, 1.50, 1.75 and 2.00 per cent, while 26.25 per cent loss was observed in *A. calamus* 0.25 per cent.

Table 4 : Effect of *Acorus calamus* rhizome powder on oviposition, hatchability and adult emergence of *Callosobruchus maculatus*

S. No.	Treatments	No. of eggs laid *	Hatchability % **	No. of adults emerged *	%Seed weight loss after 45 days **
1	*A. calamus* 0.25 %	19.00 (1.29)d	55.73 (48.88)bc	2.00 (0.46)b	1.15 (5.99)c
2	*A. calamus* 0.50 %	22.33 (1.29)d	52.46 (46.99)bc	0.00 (0.00)a	0.51 (3.53)c
3	*A. calamus* 0.75 %	12.66 (1.13)d	53.04 (47.32)bc	0.00 (0.00)a	0.15 (1.63)b
4	*A. calamus* 1.00%	12.00 (1.11)c	55.47 (48.72)bc	0.00 (0.00)a	0.00 (0.52)a
5	*A. calamus* 1.25%	9.33 (1.01)bc	44.79 (42.55)b	0.00 (0.00)a	0.00 (0.52)a
6	*A. calamus* 1.50 %	10.66 (1.06)bc	58.78 (31.76)a	0.00 (0.00)a	0.00 (0.52)a
7	*A. calamus* 1.75 %	9.00 (0.99)ab	58.73 (50.62)bc	0.00 (0.00)a	0.00 (0.52)a
8	*A. calamus* 2.00%	7.00 (0.90)a	56.87 (49.59)bc	0.00 (0.00)a	0.00 (0.52)a
9	Control	265.00 (2.42)e	75.21 (60.86)d	148.66 (2.17)c	26.25 (30.78)d

*Figures in parentheses are transformed logarithmic values ** Values in parentheses are transformed arcsine values In a column means followed by same letter(s) are not significantly different (p=0.05) by DMRT Each value is a mean of 3 replications

A. calamus 10 D (talc as filler) (Ac10DT) at 0.50, 1.00, 1.50, 2.00, 2.50,3.00,3.50 and 4.00 per cent treated black gram seeds inhibited oviposition, hatchability and adult emergence of *C. maculatus* and seed weight loss (Table 5). Ac10DT 4.00 per cent out ranked all other treatments in inhibiting the egg laying with 16 eggs followed by Ac10DT 3.50 per cent (24.33Nos) and 3.00 per cent (26.33Nos). Among the different concentrations of Ac10DT, maximum eggs were laid by female in Ac10DT 0.50 per cent (67.66Nos).

Ac10DT 4.00 per cent registered 68.00 per cent hatchability followed by Ac10DT 3.50 per cent (68.76%) and Ac10DT 3.00 per cent (70.78%) and the above treatments were on par with each other. Few adult beetles were emerged from Ac10DT 4.00 per cent (7.66Nos) followed by Ac10DT 3.50 per cent (11.66Nos), Ac10DT 3.00 per cent (15.00Nos) and Ac10DT 2.50 per cent (16.66Nos). In talc 4 per cent and untreated control 81.00 and 124.66 beetles were emerged respectively. Pandey *et al.* (1976) found that *A. calamus* rhizome powder 1% (w/w) admixed with mungbean inhibited the adult emergence of *C. maculatus*. The lowest percentage of seed weight loss was observed in Ac10DT 4.00 per cent (0.10%) followed by Ac10DT 3.50 per cent (0.30%) and Ac10DT 3.00 per cent (0.30% and these treatments were on par statistically. Germination percentage was maximum in Ac10DT 4.00 per cent (93.33%) which was significantly superior to all other treatments. Pandey *et al.* (1976) reported that 100 parts of *C. arietinum* mixed with 2 parts of sweet flag powder gave good protection against *C. chinensis* attack.

Table 5 : Effect of *Acorus calamus* 10D (talc as filler) (Ac 10 DT) on oviposition, hatchability and adult emergence of *Callosobruchus maculatus*

S. No.	Treatments	No. of eggs laid *	Hatchability % **	No. of adults emerged *	%Seed weight loss after 45 days **	Seed germination after 90 days**%
1	Ac 10 DT 0.50 %	67.66 (1.83)c	82.77 (65.49)bc	46.33 (1.66)e	12.28 (20.50)d	28.00 (31.92)cd
2	Ac 10 DT 1.00 %	58.66 (1.76)c	86.75 (68.93)c	42.66 (1.62)c	11.83 (20.11)d	33.33 (35.25)cd
3	Ac 10 DT 1.50 %	53.66 (1.72)c	80.79 (64.06)b	36.00 (1.55)e	8.66 (16.96)c	34.44 (30.06)d
4	Ac 10 DT 2.00 %	37.00 (1.56)b	81.29 (64.39)b	22.00 (1.33)d	2.21 (8.29)b	38.66 (38.44)cd
5	Ac 10 DT 2.50 %	27.00 (1.42)b	81.28 (63.88)b	16.66 (1.21)cd	1.85 (7.33)b	44.00 (41.54)c
6	Ac 10 DT 3.00 %	26.33 (1.41)b	70.78 (57.27)a	15.00 (1.17)cd	0.30 (3.04)a	66.66 (54.77)b
7	Ac 10 DT 3.50 %	24.33 (1.38)b	68.76 (56.08)a	11.66 (1.17)bc	0.15 (2.19)a	81.33 (64.43)a
8	Ac 10 DT 4.00 %	16.00 (1.19)a	68.00 (56.04)a	7.66 (1.06)b	0.10 (1.77)a	93.33 (75.20)a
9	Activated clay 1.00%	55.00 (1.73)c	81.16 (64.37)b	37.00 (0.87)a	11.01 (19.30)d	38.66 (38.43)cd
10	Talc 4.00%	114.00 (2.05)d	80.65 (63.92)b	81.00 (1.90)f	12.00 (20.26)c	26.66 (31.07)cd
11	Control	204.00 (2.30)e	78.89 (62.65)b	124.66 (2.09)g	18.30 (25.32)e	10.66 (18.98)e

*Figures in parentheses are transformed logarithmic values
** Figures in parentheses are transformed arcsine values
In a column means followed by same letter(s) are not significantly different (p=0.05) by DMRT, Each value is a mean of 3 replications

A. calamus 10 D (fly ash as a filler) (Ac10DF) same doses as (Ac10DT) was tested for oviposition, hatchability and adult emergence, weight loss and germination were recorded (Table 6). Ac10DF 4.00 per cent had less number of eggs and adult emergence. In Ac10DF 4.00 per cent 18 eggs were laid followed by Ac10DF 3.50 per cent (28.66Nos) and these treatments were on par with each other. The hatchability per cent was minimum in Ac10DF 4.00 per cent (61.26 %) followed by Ac10DF 1.00 per cent (62.25%) and maximum in Ac10DF 3.00 per cent (76.59%) and the above treatments were significantly different from control. Few adult beetles were emerged from black gram seeds treated with Ac10DF 4.00 per cent (6.66Nos) followed by Ac10DF 3.50 per cent (10.33Nos) and these treatments were on par with each other.

Table 6 : Effect of *Acorus calamus*10D (Fly ash as filler) (Ac 10 DF) on oviposition, hatchability and adult emergence of *Callosobruchus maculatus*

S. No.	Treatments	No. of eggs laid *	Hatchability % **	No. of adults emerged *	%Seed weight loss after 45 days **	Seed germination after 90 days**%
1	A_c 10 DF 0.50 %	58.33 (1.76)c	73.23 (59.88)bcd	29.33 (1.45)de	8.75 (17.17)de	30.66 (33.61)g
2	Ac 10 DF 1.00 %	69.00 (1.65)bc	62.25 (52.15)a	29.33 (1.46)de	9.15 (17.59)e	33.33 (35.25)e
3	Ac 10 DF 1.50 %	59.66 (1.77)c	68.69 (56.03)abc	26.33 (1.41)cde	6.58 (14.83)c	37.33 (37.65)e
4	Ac 10 DF 2.00 %	58.00 (1.76)c	62.78 (52.41)a	22.33 (1.34)cd	1.91 (7.64)b	48.00 43.85)d
5	Ac 10 DF 2.50 %	51.33 (1.71)c	63.14 (52.73)ab	19.66 (1.29)bc	2.55 (9.01)b	45.33 (42.31)d
6	Ac 10 DF 3.00 %	43.00 (1.63)bc	76.59 (61.14)cd	18.33 (1.26)b	1.81 (7.53)b	65.33 (53.93)c
7	Ac 10 DF 3.50 %	28.66 (1.45)ab	65.21 (53.92)ab	10.33 (0.98)a	2.75 (8.02)b	74.66 (59.88)b
8	Ac 10 DF 4.00 %	18.00 (1.25)a	61.26 (51.52)a	6.66 (0.81)a	0.11 (4.25)a	89.33 (71.01)a
9	Activated clay 1.00 %	41.33 (1.60)bc	84.63 (66.92)d	26.66 (1.39)cde	7.14 (15.69)cd	48.00 (43.85)d
10	Fly ash 4.00%	103.00 (2.01)d	84.80 (67.07)d	68.33 (1.48)e	16.18 (23.70)f	20.00 (26.49)f
11	Control	188.00 (2.27)e	77.60 (61.87)cd	119.66 (2.07)f	21.93 (27.92)g	9.15 (17.98)g

*Figures in parentheses are transformed logarithmic values
** Figures in parentheses are transformed arcsine values
In a column means followed by same letter(s) are not significantly different (p=0.05) by DMRT
Each value is a mean of 3 replications

Seed weight loss was 0.11 per cent in Ac10DF 4.00per cent which was significantly different from all other treatments. Germination percentage was maximum in Ac10DF 4.00 per cent (89.33%) followed by Ac10DF 3.50 per cent (74.66%) and minimum in Ac10DF 0.50 per cent (30.66%). In activated clay 1.00 per cent, fly ash 4 per cent and control, 48.00, 20.00 and 9.15 per cent germination was observed. Similarly, no reduction in germination potential of green gram treated with *A. calamus* rhizome powder was observed by Chiranjeevi (1991).

References

Adams,J.M. and G.G.M.Schulten.1978. Post harvest grain loss assessment methods. American Assn. of cereal chemists.,St. Paul,Minnesota,USA.193 pp135.

Chiranjeevi. C.H. 1991. Efficacy of some indigenous plant materials and ashes on the percentage of damaged grains, percentage of protection and viability of green gram seed infested by pulse beetle, *Callosobruchus chinensis* (L).*Bull. Grain Tech.*, 29 (2) : 84-88.

Credland, P.F. and A.W. Wright.1989. Factors affecting female fecundity in the cowpea seed beetle, *Callosobruchus maculatus* (Coleoptera : Bruchidae). *J. Stored Prod. Res.*, 25(3): 125-139.

Golob, P. 1984. Improvements in maize storage for the smallholder farmer. *Trop. Stored Prod. Info.*, . 50 :14-19.

Golob, P. and D.J. Webley. 1980. The use of plants and minerals as traditional protectants of stored products. *Tropical Products Institute* G 138. Now Post harvest pest and quality section, Natural Resources Institute, Chatham, United Kingdom.32 pp

Igantowicz, S. and B. Wesolowska. 1996. Repellency of powdered plant materials of Indian neem tree, the Labrador tea and sweet flag, to some stored product pests. *Polskie Pismo Entomologiczne*, 65 (1): 61-67.

Ketkar, C.M. 1987. Use of tree derived non edible oils as surface protectants for stored legumes against *Callosobruchus maculatus* and *C. chinensis*. In: pp. 535-542. Schmutterer, H. & Ascher, K.R.S., Eds, *Proc. IIIrd Intl. Neem Conf.* July 1986, Nairobi, Kenya.

Khan, M.I. and M.M. Borle.1985. Efficacy of some safer grain protectants against the pulse beetles,*Callosobruchus chinensis* (L). infesting stored bengal gram (*Cicer arietinum* L.). *Punjabrao Krishi Vidhayapeeth Research Journal*, 9 (1): 53-55.

Mohapatra, G.K. and G.P. Gupta. 1998. Pesticide induced resurgence. *Pestology*, 22 (12): 14-20.

Pandey, N.D., S.R. Singh and G.C. Tiwari. 1976. Use of some plant powders, oil and extracts as protectants against pulse beetle, *Callosobruchus chinensis*. Linn. *Ind. J. Ent..*,38 (2): 110-113

Rao, P.K., M.A. Aleem K.C. Chitra and A. Mani. 1993. Efficacy of some botanicals and ash against pulse beetle, *Callosobruchus chinensis* (Linnaeus). *Botanical Pesticides in Integrated Pest Management*. 282-287 p.

Stoll, G. 1988. Natural plant protection in the tropics. AGRECOL, Margraf Publishers Scientific Books, Weikersheim, Germany.

Senthil kumar C.M., S.Ravi kumar T. Swaminathan U.R. Maheswari and P.Narayanasamy.1988. Fly ash solves problems of pest in soybean, *Kisan World*, 12-13

Jeyarajan Nelson, S., N. Murugesan, V.K.R. Sathiyanandam, Y.S. Johnson Thangaraj Edward and M.Muthusamy.2003.A new botanical insecticide for managing rice bug. *Intl. Rice Res. News*, 28 (1): 44.

Shivanna, S., S.Lingappa, and B. Patil. 1994. Effectiveness of selected plant materials as protectants against pulse beetle, *Callosobruchus chinensis* (Linn.) during storage of red gram. *Kar. J. Agri. Sci., 7*(3): 285-290.

— xxx —

22

Screening of Different Insecticides as Jute Fabric Treatment to Protect Wheat Grains Against Lesser Grain Borer, *Rhizopertha dominica* (Fab.) Coleoptera: Bostrichidae

JHALA, R. C. AND R. H. KHER
Department of Entomology, B. A. College of Agriculture, Anand Agricultural University, Anand – 388 110

SUMMARY

Insecticide application as impregnation in jute bag was found more effective than spray on jute bag. B-twill jute bag was found more effective than hessian jute bag for insecticide spray or impregnation. Overall, based on adult mortality, half-life and gross persistency values, deltamethrin was found effective and long lasting grain protectants of wheat against *R. dominica* when sprayed/impregnated on/in jute bags as prophylactic treatment. Cypermethrin, fipronil and fenvalerate were next to deltamethrin. Spark, malathion, endosulfan and Nurelle-D were least effective.

Key words: Gross Persistency, Hessian Cloth, Jute Bag Impregnation.

1. INTRODUCTION

Bag storage is largely in vogue in farmer, government and trade godowns. The points in favour of bag storage are: (a) that each bag constitutes a definite unit which as such can be brought in, sent out, purchased or sold, (b) that under present Indian conditions it is easier to load and unload grain packed in bags, (c) that bags can be nearly arranged in stacks of different sizes, (d) that if any individual bag is noticed to be infested, it can be easily segregated and treated, (e) that the surface of the bag being practically exposed to atmosphere, the problem of sweating is less etc. However, storage losses due to insects in the food grains stored in bags are reported to be comparatively higher than in bulk storage (Girish *et al.*, 1974). A kernel damage of 2.7 to 66.22 per cent in wheat stored in bags was reported by Pingale (1955). To prevent new infestation or to disinfect old gunny bag, application of insecticides either as spray on the surface or as impregnation in the bags are recommended. However, a very limited work (Yadav and Singh, 1994; Yadav, 1997; Mahla, 2001; Pathak *et al.*, 2002 and Patil *et al.*, 2004) on grain protecting efficacy of different insecticides in the form of spray or impregnation on/in jute bags is carried out in last 15 years. The insecticides like fipronil, endosulfan, Spark and Nurelle-D are not evaluated at all for the above purpose. The information on suitability of jute bags prepared from two types of jute fabric (Hessian and B-twill) for impregnation or surface application of insecticides is also not available.

2. MATERIALS AND METHODS

2.1 Maintenance of *R. dominica* culture

For initiation of culture, required number of adults of *Rhizopertha dominica* L. were collected from Krishna mill, Anand-Sojitra Road, Anand. About 100-250 adults were introduced in plastic jars (20 cm Height and 14 cm Diameter) containing 250 g wheat grains (cv. Lok-1), previously sterilized at 55° C for 4 hours in oven. The jar was covered tightly with muslin cloth affixed with rubber band to prevent the adults from escaping. Such 5 jars containing wheat grains and adults of *R. dominica* were maintained in the laboratory for development and growth of the population. The damaged grains were replaced with undamaged and sterilized grains in each jar at monthly interval. The culture was maintained in the laboratory throughout the experimental period. The adults of *R. dominica* obtained from the laboratory culture were used for further studies on screening of different insecticides as jute fabric treatment to protect wheat grains.

2.2. Treatment Application

Ten different insecticides [Deltamethrin, Cypermethrin, Fipronil, Fenvalerate, Fenpropathrin, Chlorpyriphos, Endosulfan, Nurelle-D (Chlorpyriphos 50% + Cypermethrin 5%) and Spark (Triazophos 35% + Deltamethrin 1%) and malathion; all at 0.025 per cent concentration in water] were evaluated for their comparative efficacy in the form of spray application on jute bag and impregnation in jute bag to protect wheat grains against lesser grain borer, *Rhizopertha dominica* Fabricius. Simultaneously, two types of jute bags *viz.*, Hessian cloth (11 x 10 meshes) and B-twill cloth (6 x 7 mesh) were also evaluated for their grain protecting effectiveness after spray / impregnation of above insecticides. For the purpose, an experiment was carried out during April 2004 to October 2004 in laboratory condition following Completely Randomized Design with 3 factors [Method of application] (spraying and impregnation) as first factor, type of bag (Hessian cloth bag and B-twill cloth bag) as second factor and insecticides as third factor] and 3 repetitions. For each treatment, 12 sackcloth (6 of Hessian cloth and 6 of B-twill cloth) each of 10 x 8 cm size were prepared. Out of 12 sackcloth under each insecticide treatment, 6 sackcloth (3 Hessian cloth bags and 3 B-twill cloth bags) were filled up with previously sterilized 200 g of wheat grain and mouth of each bag was closed properly using jute fiber thread. All the 6 sacklet were sprayed with respective insecticide using hand sprayer. For each sacklet about 8-10 ml spray fluid of insecticide was required for uniform spray application. Under control treatment the sackcloth were sprayed with water without insecticides. The sprayed sackcloth were dried under fan in well aerated room for 24 hours. Out of 12 sackcloth under each treatment, 6 sackcloth (3 Hessian cloth sackcloth and 3 B-twill cloth sackcloth) were impregnated with respective insecticides. For impregnation of insecticide in jute sackcloth, all the 6 empty sackcloth were socked in 1 liter solution of respective insecticides in water for about 10 minutes and then dried under fan in well aerated room for 24 hours. Under control treatment the sackcloth were soaked in water without insecticides. After impregnation, each sackcloth was filled up with previously sterilized 200 g grains of wheat and mouth was closed properly using jute fiber thread.

2.3. Evaluation Based on Adult Mortality

The jute sackcloth filled up with wheat grains and sprayed/ impregnated with insecticides as described above were placed individually in plastic bowl (Diameter:16 cm, Height:7.0 cm) with netted lid. Twenty adults (five to ten days old) of *R. dominica* obtained from laboratory culture were released on sackcloth placed inside the bowl and the bowl was closed with lid. The observations on number of dead adults (inside and outside the sacklet) out of total adults were made 7 days after release of adults. The experiment was repeated at 15 day interval using same sackcloth refilled with fresh and previously sterilized 200 g wheat grain. The data on per cent adult mortality were computed, corrected using Abbott's formula (Abbott, 1925) and subjected to ANOVA after transforming them to arc sin. The data were also pooled over periods.

2.4 Evaluation based on Half-life and Gross persistency

The insecticides used as spray/impregnation on/in jute bags were also evaluated for their grain protecting efficacy against *R. dominica* based on mortality half-life and gross persistency. Mortality half-life is the time in days required to reduce the mortality to half of its initial mortality. It was worked out using following formula: Mortality half-life (M½) = Log2 / K1, where, K1 = Slope of the exponential regression of logarithms of number of adults died (after the addition of 1 to forestall the calculation of Log 2) out of 20 adults against days passed. The mortality half-lives in days were worked out repetition wise for each treatment of oil. The data on half-lives were subjected to ANOVA to see the significant difference among various oils.

The gross persistency of different insecticides tested as mortality of adults of *R. dominica* was studied as per the formula of Pawar and Yadav, (1980)

The purpose, the data on corrected per cent mortalities worked out at 15 days interval were used and the gross persistency data were calculated repetition wise for each insecticide treatment. The data on gross persistency were subjected to ANOVA to see the significant difference among various treatments.

3. RESULTS AND DISCUSSION

3.1. Evaluation Based on Adult Mortality

The periodical data on adult mortality depicted graphically in figure 1 further revealed that up to 30 days, all the insecticides could maintain the adult mortality above 90 per cent. Fipronil, cypermethrin and fenvalerate could maintain the adult mortality above 90 % up to 45 days only. While, deltamethrin recorded the adult mortality above

90 % up to 75 days. Above 50% adult mortality was maintained up to 135 days only in case of deltamethrin and cypermethrin, while it was up to 105 days in case of fipronil, fenvalerate and fenpropathrin; up to 90 days in case of chlorpyriphos and endosulfan and only up to 75 days in case of malathion, Nurelle-D and Spark. After 150 days, the adult mortality remained below 50 % in all the insecticides. Among two types of method of insecticide application on/in jute bag, impregnation remained significantly more effective than spray application throughout the experimental period. Further, up to 60 days of insecticide application, there was non-significant difference between two types of jute bag. After 75 days of insecticide application, B-twill jute bag was found significantly better than hessian jute bag and its superiority was maintained up to 135 days. After 135 days, there was non-significant difference between the two types of jute bag.

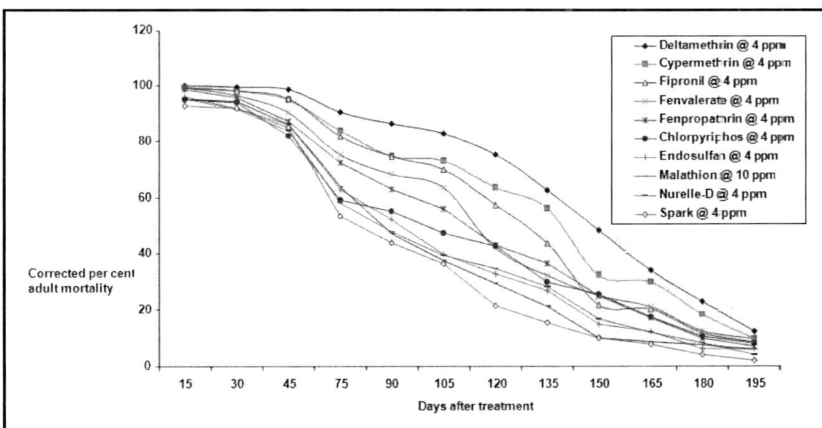

Fig. 1. Mean adult mortality of R. dominica on/in jute bags sprayed/impregnated with different insecticides.

The analysis of data on adult mortality of *R. dominica* pooled over period (Table 1) revealed that among two types of method of insecticide application, impregnation recording significantly higher mean adult mortality was more effective than spray application. However, the performance of methods was not consistent at different time interval. Among two types of jute bag, B-twill recording significantly higher mean adult mortality was found more effective than hessian jute bag. Performance of type of bags was consistent over period. Performance of methods of insecticides application varied with type of bags or vice versa and this variation was consistent over periods. The performance of insecticides varied with type of methods but not with type of bags.

The performance of insecticides with respect to type of bags was consistent over periods, while with respect to type of methods, it was inconsistent over periods. The source 'period' of ANOVA was also significant indicating that there was significant difference among mean mortalities obtained at different period which was because of decrease in efficacy of all insecticides as period of storage increased. Similarly, there was an inconsistency in performance of various insecticides at different periods of storage. The order of efficacy of various insecticides based on pooled data on adult mortality (given in bracket after each insecticide) was deltamethrin (75.24 %) > cypermethrin (66.62 %) > fipronil (62.23 %) > fenvalerate (57.54%) > fenpropathrin (54.46 %) > chlorpyriphos (49.58 %) > endosulfan (45.69 %) > Nurelle-D (45.00 %) > malathion (43.82 %) > Spark (38.92 %). Deltamethrin was found most effective and significantly superior than other insecticides. Cypermethrin followed by fipronil were next to deltamethrin in effectiveness and significantly superior to rest of insecticides. Fenvalerate and fenpropathrin were at par with each other. Similarly, endosulfan, Nurelle-D and malathion were at par with each other. Spark was found to be least effective and differed significantly from other insecticides.

Various insecticides at different doses have been evaluated as surface treatment on jute bags by different scientists. Malathion at 100 & 130 mg/Ft^2 caused knock down of *T. castaneum* for 4 weeks (McFarlane, 1961). The per cent adult mortality of *R. dominica* on jute bags sprayed with malathion (0.538 to 2.152 g/m^2) ranged between 38 to 93 % at 1 hour after treatment and decreased to 1 to 9 % after 30 days of treatment (Girish *et al.*, 1973). Due to malathion deposit at 81.3 and 41.8 mg/m^2 on jute bag, the adult mortality of *R. dominica,* which was 70 and 100 % after 1 day of treatment, reduced to zero after 20 and 10 days of treatment, respectively (Pawar and Yadav, 1980). Malathion applied at 0.15 g/m^2 on stack of gunny bag resulted into 100 % mortality of *R. dominica* adults up to 30 days after treatment (Doharey *et al.*, 1981). The average per cent mortality of *R. dominica* on jute bag surface applied with malathion (0.15 g a.i./m^2) was 70 to 75 % after 3 days of treatment, which increased to 90 to 92.5 % after 15 days of treatment (Ashok kumar *et al.*, 1982). Malathion at 0.15 g a.i./m^2 caused 70 % adult mortality of *R. dominica* on jute bag up to 15 days of treatment, which reduced to 50 % after 30 days of treatment (Lal *et al.*, 1988). At 0.15 g/m^2, it also gave cent per cent mortality of *R. dominica* after 15 day of treatment on stacks of wheat (Pandey *et al.*,

1988). At 20 mg/m^2 deposit on stacks of wheat, it resulted into total kill of *R. dominica* after 50 day of application (Yadav, 1988). At 150 mg/m^2 on stack of grain bag, it also gave cent per cent kill of *R. dominica* up to 15 days after treatment (Lal *et al.*, 1989). Fenvalerate at 1 and 1.5 g a.i./m^2 on jute bag gave 100 and 12-18 % mortality of *T. castaneum* up to 30 and 90 days, respectively during storage period (Sarkar *et al.*, 1984). Deltamethrin and cypermethrin at 20 mg/m^2 maintain more than 50 % adult mortality of *C. maculatus* and *C. chinensis* on jute surface up to 6 months (Rahman and Yadav, 1985). Both insecticides also gave complete kill of *C. maculatus* after 15 days and of *C. chinensis* after 1 month on jute bag (Yadav and Jha, 1985). Deltamethrin at 20 mg/m^2 deposit on jute bag gave total kill of *R. dominica* up to 120 days (Yadav, 1986), while at 30 mg a.i./m^2 caused 100 and 70 % adult mortality of *R. dominica* on jute bag up to 15 day of treatment which reduced to 70 and 50 % after 30 days of treatment, respectively (Lal *et al.*, 1988). Deltamethrin is also reported to give 90 % mortality of bruchids up to 180 days (Jain and Yadav, 1989). At 30 mg/ m^2 on stack of grain bag, deltamethrin resulted into cent per cent kill of *R. dominica* up to 135 days after treatment (Lal *et al.*, 1989). During present investigation deltamethrin and cypermethrin at 0.025 % as spray application on jute bag are found most effective insecticides and remained the adult mortality more than 90 % up to 75 and 45 days, respectively.

Various insecticides at different doses have been evaluated as impregnation in jute bags by different scientists. Malathion impregnated in jute bag at 1 and 1.5% gave total kill of *R. dominica* up to 15 and 16 days, respectively (Rai *et al.*, 1985a & b). Cypermethrin and deltamethrin, both at 0.0125 % were equally effective, but significantly more effective than fenvalerate (0.0125 %) which was in turn more effective than malathion (0.1 %) in preventing the damage of wheat grain stored at farmer's level (Ramzan *et al.*, 1987). Cypermethrin, deltamethrin, fenpropathrin and fenvalerate at 0.0125 % proved significantly better than malathion in giving protection against insect pests in stored wheat (Ramzan and Chahal, 1987). Chlorpyriphos (0.1%); fenvalerate and cypermethrin (0.01 and 0.025 %); and deltamethrin (0.005 %) as impregnation in gunny bags provided effective control of *R. dominica* up to 6 months in stored wheat grains. Malathion (0.1%) was less effective than synthetic pyrethroids (Mahla, 2001). As impregnation in gunny bag, deltamethrin (0.0125 %) was found more effective than malathion (0.5 %) in controlling *R.*

Table 1: Per cent adult mortality of *R. dominica* on/in jute bags sprayed/impregnated with different insecticides and filled with wheat grains (Pooled over periods)

Insecticides (All at 0.025%)	Corrected per cent adult mortality Methods of insecticide application and type of jute bags						Mean
	Spray			Impregnation			
	Hessian cloth	B-twill cloth	Mean	Hessian cloth	B-twill cloth	Mean	
Deltamethrin	50.96 (60.33)	50.84 (60.12)	50.90 (60.22)	66.86 (84.56)	71.97 (90.42)	69.42 (87.64)	60.16 a (75.24)
Cypermethrin	47.38 (54.15)	47.19 (53.82)	47.29 (53.99)	60.96 (76.44)	63.33 (79.85)	62.14 (78.16)	54.71 b (66.62)
Fipronil	43.79 (47.89)	44.54 (49.20)	44.17 (48.55)	57.34 (70.88)	62.67 (78.92)	60.00 (75.00)	52.08 c (62.23)
Fenvalerate	42.41 (45.48)	42.27 (45.28)	42.34 (45.36)	55.84 (68.47)	56.86 (70.11)	56.35 (69.29)	49.34 d (57.54)
Fenpropathrin	44.35 (48.86)	44.11 (48.45)	44.23 (48.65)	47.09 (53.64)	54.69 (66.59)	50.89 (60.21)	47.56 d (54.46)
Chlorpyriphos	36.96 (36.15)	36.28 (35.01)	36.62 (35.58)	54.80 (66.77)	50.99 (60.38)	52.89 (63.60)	44.76 e (49.58)
Endosulfan	34.09 (31.41)	34.30 (31.76)	34.19 (31.58)	47.26 (53.94)	54.47 (66.23)	50.87 (60.17)	42.53 f (45.69)
Malathion	36.81 (35.90)	37.11 (36.40)	36.96 (36.15)	44.77 (49.60)	47.10 (53.66)	45.94 (51.64)	41.45 f (43.82)
Nurelle-D [Chlorpyriphos + Cypermethrin (50:5) 55 EC]	36.79 (35.87)	36.98 (36.18)	36.88 (36.02)	46.88 (53.28)	47.86 (54.98)	47.37 (54.13)	42.13 f (45.00)
Spark [Triazophos+ Deltamethrin (35:1) 36 EC]	33.75 (30.86)	34.53 (32.13)	34.14 (31.50)	43.22 (46.89)	42.91 (46.35)	43.07 (46.63)	38.60 g (38.92)

Two way table	Type of bags	Methods of application		Mean
		Spray	Impregnation	
	Hessian cloth	40.73 (42.57)	52.50 (62.94)	46.61 (52.81)
	B-twill cloth	40.82 (42.73)	55.29 (67.57)	48.05 (55.31)
	Mean	40.77 (42.64)	53.89 (65.27)	47.33 (54.06)

contd...

contd...

		Source	S. E. m ±	CD at
ANOVA		Method (M)	0.30	0.83
		Bag (B)	0.30	0.83
		Treatment (T)	0.66	1.86
		M x B	0.42	1.18
		M x T	0.93	2.63
		B x T	0.93	NS
		M x B x T	1.32	NS
	CV %		17.44	
		Period (P)	0.55	1.52
		M x P	0.77	2.14
		B x P	0.77	NS
		T x P	1.73	4.80
		M x B x P	1.09	NS
		M x T x P	2.45	6.78
		B x T x P	2.45	NS
		M x B x T x P	3.46	NS
	CV %		12.67	

Means in parentheses are retransformed values, those outside are arc sin transformed values.
Means with letter(s) in common do not differ significantly in respective column. NS: Not significant

dominica infestation on wheat seeds during 8 months of storage. Impregnation of insecticide was found better method than spray application of insecticide on jute bags for the control of *R. dominica* (Patil *et al.*, 2004).

3.2. Evaluation Based on Half-life and Gross Persistency Values

Half-life: The analysis of data on half-life in days (Table 2) revealed that the treatment difference was significant. The chronological order of insecticides based on mean half-life in days given in bracket after each insecticides was deltamethrin (103.35) > cypermethrin (84.49) > fenvalerate (76.72) > fipronil (71.60) > fenpropathrin (67.77) > chlorpyriphos (67.35) > endosulfan (58.71) > Nurelle-D (56.91) > malathion (52.36) > Spark (46.30). Deltamethrin recording highest half-life was found most long lasting grain protectant and was significantly superior to other insecticides.

Fenvalerate was at par with cypermethrin on one side while with fipronil, fenpropathrin and chlorpyriphos on another side of chronological order. Endosulfan and Nurelle-D were at par with each other as well as with chlorpyriphos and fenpropathrin on one side while with malathion on another side of chronological order of insecticides. Spark, the least lasting insecticides was at par with malathion and Nurelle-D. Thus, deltamethrin, cypermethrin and fenvalerate recording mean half-life above 75 days can be considered as most effective insecticides for surface application on jute bags. Fipronil, fenpropathrin, chlorpyriphos, endosulfan, Nurelle-D and malathion recording half-life between 50 and 75 days can be considered as moderately effective insecticide. While Spark which recorded half-life below 50 days can be considered as least effective insecticide for their efficacy as spray or impregnation in both the type of jute bag against *R. dominica*. Further, from the view point of half-life of insecticides, impregnation was significantly more effective method than spray application on jute bags. Among two types of jute bag, B-twill was significantly superior to Hessian. Performance of method varied with the types of bag or vice versa. The performance of treatments varied with type of methods but not with type of bags.

3.3. Gross Persistency

The chronological order of various insecticides based on gross persistency values given in bracket after each insecticides (Table 3) was deltamethrin (5820.70) > cypermethrin (5116.04) > fipronil

Table 2: Half-life of insecticides evaluated for their grain protecting efficacy as spray/impregnation on/in jute bags against *R. dominica* during storage of wheat grain

Insecticides (All at 0.025%)	Corrected per cent adult mortality						
	Methods of insecticide application and type of jute bags						Mean
	Spray			Impregnation			
	Hessian cloth	B-twill cloth	Mean	Hessian cloth	B-twill cloth	Mean	
Deltamethrin	48.26	55.92	52.09	142.34	166.88	154.61	103.35 a
Cypermethrin	46.81	45.58	46.19	109.28	136.30	122.79	84.49 b
Fipronil	40.71	40.35	40.53	78.81	126.56	102.68	71.60 c
Fenvalerate	39.93	39.16	39.55	114.34	113.46	113.90	76.72 bc
Fenpropathrin	42.76	41.28	42.02	77.74	109.28	93.51	67.77 cd
Chlorpyriphos	45.90	43.69	44.79	93.28	86.54	89.91	67.35 cd
Endosulfan	40.59	40.95	40.77	59.02	94.30	76.66	58.71 de
Malathion	40.48	40.70	40.59	59.95	68.32	64.13	52.36 ef
Nurelle-D [Chlorpyriphos+ Cypermethrin (50:5) 55 EC]	42.30	42.19	42.24	77.68	65.48	71.58	56.91 def
Spark [Triazophos+ Deltamethrin (35:1) 36 EC]	39.15	40.28	39.71	54.52	51.28	52.90	46.30 f

Two way table	Type of bags	Methods of application		Mean
		Spray	Impregnation	
	Hessian cloth	42.69	86.69	64.69
	B-twill cloth	43.00	101.84	72.42
	Mean	42.85	94.27	68.56

	Source	S. Em ±	CD at 5 %
	Method (M)	1.76	4.95
	Bag (B)	1.76	4.96
	Treatment (T)	3.94	11.08
	M x B	2.49	7.01
	M x T	5.57	15.67
	B x T	5.57	NS
	M x B x T	7.88	NS
	CV %	19.90	

(4570.70) > fenvalerate (4274.80) > fenpropathrin (3998.79) > chlorpyriphos (3720.74) > endosulfan (3272.54) > Nurelle-D (3190.41) > malathion (2975.36) > Spark (2592.42). The analysis of data on gross persistency indicated that a significant difference was observed among various insecticides. Deltamethrin recording higher gross persistency was significantly superior to other insecticides. Cypermethrin was next to deltamethrin but significantly superior to rest of the insecticides. Fenvalerate was at par with fipronil on one side while with fenpropathrin on other side of the chronological order. Fenpropathrin was at par with chlorpyriphos. Endosulfan, Nurelle-D and malathion were at par with each other but differed significantly from rest of the insecticides. Spark was significantly inferior to other insecticides. Further, from the view point of gross persistency values, impregnation was significantly more effective than spray application of insecticides. B-twill jute bag recorded significantly higher mean gross persistency value than hessian jute bag (Figure 26). Performance of method varied with type of bags or vice-versa. Performance of insecticides was not consistent in two types of insecticide application and in two types of bag.

Thus, based on half-life and gross persistency values, deltamethrin was found long lasting grain protectants of wheat against *R. dominica* when sprayed/impregnated on/in jute bags. Cypermethrin and fipronil were next to deltamethrin for their long lasting grain protectant efficacy. Spark, malathion, endosulfan and Nurelle-D were least effective while fipronil, fenpropathrin and chlorpyriphos were found moderately effective insecticides as surface treatment on jute bag.

The persistency of insecticides as surface application on jute bags is reported by various workers. At 81.3 and 41.8 mg/m^2 deposit of malathion on jute bag, the gross persistency values were 47.2 and 48.1, respectively based on periodical mortality of *R. dominica*. (Pawar and Yadav, 1980). Deltamethrin and cypermethrin at 20 mg/m^2 maintained more than 50 % adult mortality of *C. maculatus* and *C. chinensis* on jute surface up to 6 months. The gross persistency values were also more than 5000 after 6 months of application of above two insecticides. Deltamethrin was most toxic and persistent followed by cypermethrin and fenvalerate (Rahman and Yadav, 1985). Cypermethrin was found less stable than deltamethrin (both at mg/m^2) on jute surface when tested against *C. maculatus* and *C. chinensis* (Yadav and Jha, 1985). Deltamethrin at 20 mg/m^2 on jute surface persisted up to 120 days and gave total kill of *R. dominica* (Yadav,

Table 3 : Gross persistence of insecticides evaluated for their grain protecting efficacy as spray/ impregnation on/in jute bags against *R. dominica* during storage of wheat grain

Insecticides (All at 0.025%)	Corrected per cent adult mortality						
	Methods of insecticide application and type of jute bags						Mean
	Spray			Impregnation			
	Hessian cloth	B-twill cloth	Mean	Hessian cloth	B-twill cloth	Mean	
Deltamethrin	4096.98	4113.60	4105.29	7270.08	7802.12	7536.10	5820.70 a
Cypermethrin	3717.38	3724.04	3720.71	6293.81	6728.91	6511.36	5116.04 b
Fipronil	3137.18	3233.29	3185.24	5337.19	6575.15	5956.17	4570.70 c
Fenvalerate	2921.90	2937.57	2929.74	5437.46	5802.26	5619.86	4274.80cd
Fenpropathrin	3336.62	3350.57	3343.59	3824.56	5483.41	4653.99	3998.79de
Chlorpyriphos	2426.81	2487.48	2457.14	4670.66	5298.02	4984.34	3720.74 e
Endosulfan	2078.87	2151.99	2115.43	3679.89	5179.40	4429.64	3272.54 f
Malathion	2414.08	2471.14	2442.61	3292.49	3723.72	3508.10	2975.36 f
Nurelle-D [Chlorpyriphos Cypermethrin (50:5) 55 EC]	2479.52	2535.30	2507.46	3881.74	3888.99	3885.37	3196.41 f
Spark [Triazophos+ Deltamethrin (35:1) 36 EC]	2036.54	2104.44	2070.49	3094.22	3134.68	3114.45	2592.47 g

Two way table	Type of bags	Methods of application		Mean
		Spray	Impregnation	
	Hessian cloth	2870.66	4744.99	3807.83
	B-twill cloth	2904.87	5294.88	4099.88
	Mean	2887.77	5019.94	3953.85

	Source	S. Em ±	CD at 5 %
ANOVA	Method (M)	50.77	140.93
	Bag (B)	50.77	140.93
	Treatment (T)	111.98	315.13
	M x B	70.82	199.31
	M x T	158.36	445.66
	B x T	158.36	315.13
	M x B x T	223.95	630.26
	CV %	98.1	

1986). Deltamethrin at 30 mg/m^2 was found more effective than malathion (0.15 g/m^2) as surface treatment against *R. dominica* (Lal *et al.*, 1988). Deltamethrin sprayed at 10 and 20 mg/m^2 on jute bag protected rice grain from *R. dominica* up to 6 months (Yadav, 1988). On jute surface, the gross persistency value of deltamethrin and malathion after 180 days of aging ranged between 5575 to 8875 and 4925 to 7875, respectively against 3 species [*C. maculatus, C. chinensis* and *analis* (Fabricius)] of pulse beetles, indicating higher persistency of deltamethrin than malathion (Jain and Yadav, 1989). Deltamethrin at 30 mg/m^2 was more effective and had longer residual effect as surface treatment on jute bag against *R. dominica* as compared to malathion (150 mg/m^2) as reported by Lal *et al.* (1989). Based on gross persistency value, it was concluded by Yadav and Singh (1994) that malathion retained its full toxicity up to 60 days against *C. maculatus* at dosage 100-150 mg/m^2 on jute fabric. Deltamethrin at 20 mg/m^2 could retain its full toxicity up to 180 days against *S. oryzae* and *C. maculatus*, while chlorpyriphos at 50 mg/m^2 could do so up to 120 days against *S. oryzae* and 150 days against *C. chinensis* (Yadav and Singh, 1994). An excellent protection against *R. dominica* was obtained by Yadav (1997) with deltamethrin application on jute fabric prior to the storage of wheat grains. Deltamethrin (20 mg/m^2 was most effective, followed by chlorpyriphos (100 mg/m^2) and malathion (150 mg/m^2), as surface treatment on jute bag in controlling the pests of stored maize and rice (Pathak *et al.*, 2002). Synthetic pyrethroids are found more effective than chlorpyriphos and malathion by most of the scientists which is in accordance with the present report.

The persistency of insecticides as impregnation in jute bags is reported by different scientists. Malathion impregnated bags gave good protection for a period of 2 months against *S. oryzae* and *T. castaneum* (Pali, 1960). Jute sackcloth impregnated in 0.05 % malathion solution controlled entry of *T. castaneum* and *Lasioderma serricorne* (Fabricius) into the sackcloth up to 3 months of storage (Joshi and Kaul, 1966). Malathion in jute bag was effective against stored grain pests only for 4 weeks under Indian conditions (Quadir and Syed, 1970). Malathion impregnated in gunny bag at 100 mg/Ft^2 was significantly effective in reducing the penetration of adults of *C. chinensis* in the bag (Kuppuswamy and Subramanium, 1976). Jute bags impregnated with 0.005 % deltamethrin, 0.05 % cypermethrin and 0.05 % fenvalerate were most effective against natural infestation of *R. dominica* during 8 months of storage of wheat. Pyrethroids were significantly better than organophosphate insecticides in giving protection to wheat stored in impregnated bags (Chahal and Ramzan, 1982). Jute bags impregnated with malathion (0.5 and 1.0 %) gave total mortality up to 60 days (Rai *et al.*, 1985b). During present investigation synthetic pyrethroids are found most effective as impregnation in jute bags, which are in accordance with most the reports of earlier workers. It was further indicative from the review of available literature that fipronil, endosulfan, Nurelle-D and Spark are not evaluated for their grain protecting efficacy either as spray or impregnation in jute bag, therefore as such their efficacy could not be compared with the report of earlier workers.

References

Abbott, W.S. 1925. A method of computing the effectiveness of an insecticide. *J. Eco. Entomol.*, 18: 265-267.

Ashok kumar, G.P. Pandey, R.B. Doharey and B.K. Varma 1982. Field trials with some newer organophosphatic insecticides against insect pests of stored food grains. *Pesticides*, 16(1): 7-10.

Brower, J.H. and E.W. Tilton 1973. Weight loss of wheat infested with gamma-radiated *Sitophilus oryzae* (L.) and *Rhyzopertha dominica* (F.). *J. Stored Prod. Res.*, 9: 37-41.

Chahal, B.S. and M. Ramzan 1982. Relative efficacy of synthetic pyrethroids and some organophosphate insecticides against the larvae of Khapra beetle. *J. Res. PAU.* 19(2): 123 126.

Chapbell, A. and R.N. Sinha 1976. Damage of wheat by feeding of some stored product beetles. *J. Eco. Entomol.*, 69(1): 11-13.

Doharey, R.B., A. Kumar and B.K. Verma 1981. Results of field trials with pirimiphos methyl 50 % EC against stored grain insect pests. *Pesticides*, 15(11): 7-11.

Girish, G.K. 1978. Insect problems in storage and control techniques under Indian conditions. Paper presented at the second International working conference on Stored Product Entomology, Idban Nigeria held in September 1978.

Girish, G. K.; Goyal, R. K and Krishnamurthy, K. 1973. Pirimiphos-methyl (50 per cent EC) versus malathion (50 per cent EC)- Its efficacy and residual toxicity against some stored grain insect pests. *Bull. Grain Tech.*, 11(2): 106-112.

Girish, G.K., S.K. Jain, Ashok Kumar and N.S. Agrawal 1975. Part-V: Assessment of storage losses, Quality and pesticidal contamination in wheat available in the markets of western Uttar Pradesh, Punjab and Haryana. *Bull. Grain Tech.*, 13(1): 1-18.

Girish, G.K., B.P. Tripathi, R.P.S. Tomer and K. Krishnamurthy 1974. Studies on the ssessment of Losses. *Bull. Grain Tech.*, 12(3): 199-210.

Jain, S. and T.D. Yadav 1989. Persistence of deltamethrin, etrimfos and malathion on different storage surfaces. *Pesticides*, 23(11): 21-24.

Joshi, H.C. and C.L. Kaul 1966. Studies on the protectivity of jute bags impregnated with organic insecticides against rust red flour beetle and cigarette beetle. *Indian J. Ent.*, 27: 491-493.

Kuppuswamy, S. and T.R. Subramaniam 1976. Studies on the impregnation of gunny bags with certain organophosphorus insecticides. *Bull. Grain Tech.*, 14(1): 45-50.

Lal, S., R.B. Doharey, R.K. Goyal and D. Singh 1989. Efficacy and residual toxicity of deltamethrin flow 25 as compared to malathion 50 per cent EC against stored grain insect pests in commercial warehouses. *Bull. Grain Tech.*, 27(2): 79-83.

Lal, S., S. Shanker, A.S. Yadav, Ashok kumar and J.L. Srivastava 1988. Field trial with deltamethrin against stored grain insect pests. *Bull. Grain Tech.*, 26(3): 187-190.

Mahla, J.C. 2001. Role of gunny bags dipped in different insecticides for checking insect pests infestation to the stored wheat grains under ambient storage conditions. *Ann. of Biol.*, 17(1): 95-99.

McFarlane, J.A. 1961. *J. Food Sci. Technol.*, 7: 8-10. [Fide: Kuppuswamy, S. and T.R. ubramanium (1976). *Bull. Grain Tech.*, 14(1): 45-50].

Pali, N.S. 1960. Studies on jute bug impregnation for the control of graminivorous insects. *Indian J. Ent.*, 22(3): 211-213.

Pandey, G.P., S. Shanker and P.C. Gupta 1988. Efficacy of malathion for controlling stored grain insect pests in godowns. *Bull. Grain Tech.*, 26(3): 88-89.

Patel, M.A. 1983. The assessment of the inherent vulnerability of wheat varieties to post harvest infestation by *Trogoderma granarium* E. and *Rhizopertha dominica* F. under various ecological conditions. *Ph. D. (Agri.) Thesis*, GAU, Anand.

Pathak, K.A., A.N. Jha and J.P. Singh 2002. Efficacy of fabric treatment of jute and polypropylene bags with some insecticides on maize and paddy stored at Delhi and Meghalaya. *Shashpa*, 9(1): 61-70.

Patil, S.K., S.V. Tanpur and S.N. Mote 2004. Efficacy of gunny bag treatment on natural infestation of lesser grain borer, *Rhizopertha dominica* (Fab.) during storage of wheat. *Seed Res.*, 32(1): 72-75.

Pawar, C.S. and T.D. Yadav 1980. Persistence of organophosphorus insecticides on different surfaces. *Indian J. Ent.*, 42(4): 728-736.

Pingale, S.V. 1955. Impregnation of jute bag containers with insecticides for protecting stored food stuffs. *Indian J. Ent.*, 17(3): 295-306.

Quadir and S.H. Syed 1970. *Food Sci. Technology*, 7: 8-10. [Fide: Kuppuswamy, S. and Subramanium, T. R. 1976. *Bull. Grain Tech.*, 14(1): 45-50].

Rahman, M and T.D. Yadav 1985. Persistence of deltamethrin, cypermethrin, permethrin and fenvalerate on different surfaces. *Bull. Grain Tech.*, 23(2): 103-110.

Rai, R.S., P. Lal and P.K. Srivastava 1985a. Impregnation of jute bag with insecticides for protection of stored food grains: I Preliminary studies on the efficacy of some safe organophosphatic insecticides. *Pesticides*, 19(3): 16-18.

Rai, R.S., P. Lal and P.K. Srivastava 1985b. Impregnation of jute bag with insecticides for protection of stored food grains: II Comparative efficacy of some safe organophosphatic insecticides. *Pesticides*, 19(4): 45-48.

Ramzan, M. and B.S. Chahal 1987. Evaluation of synthetic pyrethroids for the protection of stored wheat grains against storage pests. *Int. Pest Control*, 29(2): 42-44.

Ramzan, M., B.S. Chahal, B.K. Judge and D.D. Narang 1987. Field trials on the impregnation of gunny bags with synthetic pyrethroids against storage loss of wheat due to stored grain pests. *Bull. Grain Tech.*, 25(2): 160-163.

Sarkar, D.K., S.K. Handa and S.K. Mukerjee 1984. Residual toxicity and persistence of fenvalerate on jute bags and its persistence on wheat grains. *Pesticides*, 18(5): 30-31.

Srivastava, P.K., B.P. Tripathi, G.K. Girish and K. Krishnamurthy 1973. Studies on the Assessment of Losses. *Bull. Grain Tech.*, 11(2): 129-139.

Yadav, T.D. 1986. Efficacy of deltamethrin against insect infestation in wheat stored under rural storage system. *Pesticides*, 20(10): 12-13.

Yadav, T.D. 1988. Efficacy of deltamethrin against storage insects in rice and wheat under FCI's storage system. *Pesticides*, 22(12): 39-43.

Yadav, T.D. 1997. Evaluation of deltamethrin as fabric treatment in storage of wheat seed at farm level. *Indian J. Ent.*, 59(1): 103-109.

Yadav, T.D. and A.N. Jha 1985. Persistence of deltamethrin, cypermethrin and permethrin on storage surfaces. *Pesticides*, 19(7): 28-30.

Yadav, T.D. and S. Singh 1994. Persistence toxicity and efficacy of four insecticides as jute fabric treatment to protect cereal and legume seeds. *Indian J. Ent.*, 56(2): 146-155.

— xxx —

23

Toxicity of Sulfuryl Fluoride to Stored Grain Insects

SRIRANJINI V., Y. RAJASHEKAR AND S. RAJENDRAN
Food Protectants and Infestation Control Department,
Central Food Technological Research Institute, Mysore - 570 020

SUMMARY

Toxicity tests were conducted against mixed-age cultures of the rice weevil *(Sitophilus oryzae)*, the lesser grain borer *(Rhyzopertha dominica)*, the rust-red flour beetle *(Tribolium castaneum)* and the saw-toothed grain beetle *(Oryzaephilus surinamensis)* and late larvae of Khapra beetle *(Trogoderma granarium)*. *S, oryzae* and *O. surinamensis* were highly tolerant to sulfuryl fluoride that total population extinction was achieved in 24 hr exposure at 40 mg/L dose only. In contrast, Khapra larvae and all developmental stages of *Tribolium castaneum.* were quite susceptible that 100% mortality was attained at <5 mg/L doses in 24 or 48 hr exposure. The results indicate that Khapra larva, generally to be the most tolerant pest to fumigants and *Tribolium castaneum*, a predominant stored grain pest in India, could be easily controlled by sulfuryl fluoride.

Key words: Toxicity, Fumigant, Phosphine.

1. INTRODUCTION

Phosphine and methyl bromide are the principal fumigants deployed for insect pest control in stored products including food grains world over. Several stored product insects have developed different degrees of resistance to phosphine and hence, there is a problem in effective treatment of commodities using phosphine (Subramanyam and Hagstrum, 1996). In view of its ozone depleting potential, the use of methyl bromide is banned or restricted in various countries (MBTOC, 2002). Consequently, intensive research is going on in developed and developing countries on alternative fumigants such as carbonyl sulfide, ethyl formate, ethane dinitrile, propylene oxide, and sulfuryl fluoride (Rajendran, 2001).

Sulfuryl fluoride is a non-flammable, colourless, odorless inorganic gas available in steel cylinders as a liquid (99.8%) under pressure (Table 1). Kenaga (1957) reported the fumigant properties of sulfuryl fluoride against cockroaches *(Blatella germanica* and *Periplaneta americana),* the housefly *(Musca domestica)* and selected stored product insects and Stewart (1957) demonstrated its toxicity against the dry wood termite, *Kalotermes minor*. Subsequently, the compound has been used as a structural fumigant for controlling termites and woodborers in USA (trade name: 'Vikane') since 1961, in Germany since 1990 and in China (trade name: 'Xunmienjin') since 1983. Recently, sulfuryl fluoride has been registered in USA (Trade name: 'Profume'), Canada and European countries for fumigation of food commodities and food facilities.

It is known that sulfuryl fluoride doesn't cause cancer and other health effects. It is non-corrosive (Kenaga, 1957) and has no adverse effect on electronic equipments such as computers (Bell, 2006). It has been established that sulfuryl fluoride, in view of its high vapour pressure (13,442 mm of Hg at 25°C), the penetrate into commodities including flour faster than methyl bromide (Kenaga, 1957; Bell, 2006). When compared with methyl bromide it is less sorbed by food commodities but the desorption is faster than methyl bromide. The fumigant can be used under normal atmospheric pressure as well as under vacuum (Guogan *et al.,* 1999). In USA, tolerance limits of sulfuryl fluoride residues have been established for various food commodities (Prabhakaran, 2006).

In spite of the recent developments with regard to sulfuryl fluoride as a fumigant for stored food commodities, there is lack of information on doses for 100% population extinction of different stored grain insect pests. Therefore, the toxicity of the gas to mixed-age cultures (all developmental stages) of the rust-red flour beetle, *Tribolium castaneum* Hbst., the saw-toothed grain beetle, *Oryzaephilus surinamensis* L., the rice weevil, *Sitophilus oryzae* L. the lesser grain borer, *Rhyzopertha* L. *dominica* and the Khapra beetle, *Trogoderma granarium* L. (larvae only) was determined.

Table 1 : Properties of sulfuryl fluoride in comparison with currently used fumigants

S. No.	Properties	Sulfuryl fluoride	Methyl bromide	Phosphine
1	Molecular weight	102	95	34
2	Boiling point (° C)	-55.2	3.6	-87
3	Vapour pressure at 25U C (mm of Hg)	13,442	1,824	29,260
4	Specific gravity of gas (Air=l)	3.5	3.4	1.2
5	Latent heat of vapourization (cal/g)	45.1	61.5	
6	Flammability limits (%v/v)	Nonflam mable	10-16	1.8
7	Water solubility (g/L)	0.75	13.4	0.35
8	TLV-TWA (ppm)*	5	5	0.3

* Threshold limit value-Time weighted average concentration in air.

2. MATERIALS AND METHODS

2.1. Preparation of Mixed-Age-cultures

T.castaneum was reared on whole wheat flour fortified with 5% dried yeast, *O. surinamensis* on a 5:5:1 mixture of broken wheat, rolled oats and yeast, whereas *S. oryzae, R. dominica* and *T. granarium* were reared on whole wheat. T.*granarium* and *R. dominica* cultures were maintained at 30±1°C and 70% r.h. and cultures of other insects were maintained at 25±1°C and 65±10% r.h. From these cultures, adults (1-2 weeks old) were taken for preparation of mixed-age cultures. About 300 adults of T.*castaneum* and *O. surinamensis* were released into 1 kg of respective culture medium in 2 litre glass jars. Similarly, 300 adults of *S. oryzae* and *R. dominica* were allowed to breed separately in 1 kg of wheat in 2 litre glass jars. After one week, the adults were sifted from the cultures. A series of cultures of the respective species were thus maintained continuously. The insect culture jars were kept under the rearing temperature conditions i.e. 30 ± 1°C for *R. dominica* and 25 ± 1°C for other insects. Insect cultures in six successive weeks (containing 5-6, 4-5, 3-4, 2-3, 1-2, and 0-1 week old insects) were pooled such that the pooled populations contained all developmental stages of the particular species. The pooled mixture of individual species served as mixed-age cultures for toxicity tests. For *T. granarium,* instead of mixed-age culture, only the late larvae (3 weeks old) were tested.

Rearing media containing mixed-age cultures of individual species was weighed separately in 50 g aliquots into cloth bags (20 cm x 14 cm size). In the case of *T. granarium,* late larvae (3 weeks old), 20 each, were counted into cloth bags containing 50 g wheat for testing. Cloth bags containing mixed-age cultures of individual species and Khapra larvae were placed individually in 0.85 L desiccators that served as the fumigation chambers. The desiccators were provided with holes sealed by rubber septa for injecting sulfuryl fluoride.

2.2. Fumigation

The insects were exposed to a range of doses of sulfuryl fluoride (0.3 - 30 mg/L) for 24 hr and 48 hr at 26 ±2° C. For each species, there were 4 or 5 replicates per dose with an equal number of untreated control replicates. Sulfuryl fluoride was drawn from a cylinder source (purity 99% from M/s. Linhai Limin Chemicals Ltd., China) for dosing the desiccators. Gas-tight micro syringes were used for injecting the fumigant. Before injecting sulfuryl fluoride, an equivalent volume of air was drawn from the chambers (desiccators) to avoid development of any positive pressure inside.

2.3. Post-Fumigation Operations

At the end of exposure, the test insect bags were taken out of the desiccators. The contents of the bags were transferred to individual bottles (12 cm x 5 cm size) and kept at the rearing temperature and humidity conditions for 8 weeks. The insects, which emerged from wheat *(S. oryzae* and *R. dominica)* or survived as adults *(T. castaneum* and *O. surinamensis)* in their respective media were checked at weekly intervals for 8 weeks. Similarly, counts were made in untreated control batches every week. In the case of *Trogoderma granarium* larvae, mortality was assessed 10 days post-treatment. Percent kill was determined by taking the survival/emergence in the controls as 100%.

3. RESULTS AND DISCUSSION

Sulfuryl fluoride was highly toxic to *T.castaneum* and *T. granarium* that 100% mortality was recorded at the dose of 5 mg/L in 24 hr exposure itself (Table 2). *S. oryzae* and *O. surinamensis* were, however, tolerant to the fumigant that 100% mortality of the insects was achieved only at the higher dose of 40 mg/L in 24 exposures. Generally with the extended exposure period of 48 hr, increased mortality was observed in all the five species (Table 3). The order of susceptibility of the tested species did not alter with the increased exposure period. The susceptibility of the insects with 24 or 48 hr exposure was in the following decreasing order: *T. castaneum* >*T. granarium* > *R.. dominica* > *S. oryzae* > *O. surinamensis*. In untreated controls, there was no natural mortality in Khapra larvae. The average number of survivors in untreated control batches of other species varied from 49.2 *(T. castaneum)* to 576 *(S. oryzae)* in 24 hr exposure and 48.8 *(T. castaneum)* to 522 *(S. oryzae)* in 48 hr exposure.

Table 2 : Toxicity of sulfuryi fluoride to mixed-age cultures of stored grain insects at 26 ± 2 C in 24 hr exposure

Dose (mg/L)	Mortality (% ±SD) achieved *				
	Tribolium castaneum	*Trogoderma ... granarium*	*Rhyzopertha dominica*	*Sitophilus oryzae*	*Oryzaephilus surinamensis*
0.3	-	-	28.9±16.7	37.1±4.7	-
0.6	14.5±12.2	9.0±4.2	70.2±9.5	59.6±4.5	-
1.3	22.3±16.4	25.0±3.5	87.6±11.9	72.4±3.0	-
2.5	96.3±4.8	67.0±9.7	83.2±3.7	72.4±4.0	65.9±8.8
5.0	100±0	100±0	90.4±4.0	81.5±2.8	73.7±5.4
7,5			98.1±2.0	NT	NT
10				82.9±2.5	75.5±6.0
15				NT	91.2±4.0
20				93.0±1.0	92.7±2.1
30				98.6±0.8	97.2±0.6
40				100±0	100±0
Control*	49.2±7.6	20±0	72.6±9.0	576±4.0	79.2±6.4

* There were 4 or 5 replicates per dose and in untreated controls.

† Late larvae, 20 per replicate in 50 g wheat tested; for other insects, 50 g infested media per replicate tested.

• Number of survivors per replicate given. NT= Not tested

Table 3 : Toxicity of sulfuryl fluoride to mixed-age cultures of stored grain insects at 26±2° C in 48 hr exposure

Dose (mg/L)	Mortality (% ±SD) achieved *				
	Tribolium castaneum	*Trogoderma ... granarium*	*Rhyzopertha dominica*	*Sitophilus oryzae*	*Oryzaephilus surinamensis*
0.3	-	-	79.5±7.3	48.7±11.4	-
0.6	35.6±7.3	14.0±6.5	84.2±5.8	68.0±5.6	-
1.3	95.1±3.1	42.0±4.5	88.4±6.3	74.9±2.2	-
2.5	100±0	96.0±5.5	94.2±2.4	76.6±1.9	87.5±5.7
5.0		100±0	99.2±0.8	90.5±1.6	91.2±3.5
7.5			100±0	NT	NT
10				93.6±0.9	98.0±2.2
15				NT	98.3±1.2
20				99.8±0.2	99.3±1.0
30				100±0	100±0
Control	48.8±4.3	20±0	76±19.2	522±32.0	59.4±9.3

* There were 4 or 5 replicates per dose and in untreated controls.
... Late larvae, 20 per replicate in 50 g wheat tested; for other insects, 50 g infested media per replicate tested.
• Number of survivors per replicate given. NT= Not tested

In tests against all developmental stages of stored product insects such as T.*castaneum, Tribolium confusum, Ephestia kuehniella, S. granarius, Tenebrio molitor* and *Liposcelis bostrichophila,* Bell *et al.* (2002) observed that sulfuryl fluoride was effective at a concentration-time (CT) product of 500 g hr m^3 at 30° C, or 1000 g hr m^3 (= 42 g/m^3 x 24 hr or 21 g/m^3 x 48 hr) at 25° C. In the present study, for 100% population extinction of *O. surinamensis* the target CT product was high at 1440 g lir/m^3 (30 g/m^3 x 48 hr). This may be due to exceptionally high tolerance of the eggs of *O. surinamensis.* There is no previous report about the toxicity of sutfuryl fluoride to *O. surinamensis.*

In the present study, *T.castaneum* was relatively susceptible probably due to less tolerance of their eggs. In the case of *T.granarium,* only larval stage was tested. The larvae due to diapause, are tolerant to fumigants. However, sulfuryl fluoride, unlike other fumigants could cause mortality of the larvae at relatively low doses.

Desmarchelier (1994) noted that the immature stages of *S. oryzae* were highly tolerant to carbonyl sulfide fumigant also. At the dosage of 60 mg/L for 24 hr at 25° C he could achieve 99.8% mortality and at 40 mg/L for 48 hr, however, 100% mortality was achieved. Evidently, the present study shows sulfuryl fluoride is more toxic than carbonyl sulfide. In laboratory tests with the mixed-age cultures of phosphine-resistant field strain of *S. oryzae* in 7-day exposures, Rajendran *et al.* (2000) determined an effective phosphine concentration of 0.8 mg/L (CT product 0.8 g/m^3 x 168 hr - 134 g hr/ m^3). Therefore, sulfuryl fluoride is relatively less toxic than phosphine. However, unlike phosphine, sulfuryl fluoride is rapidly toxic to insect pests that 24-48 hr exposure is adequate. Kenaga (1957) demonstrated that sulfuryl fluoride is more toxic than methyl bromide to stored product insects.

The study shows that sulfuryl fluoride has the advantage of its high toxicity to Khapra larvae, which is generally tolerant to other fumigants and *T.castaneum,* a serious pest of stored grain in warehouses in India and in flourmills. Like phosphine, for sulfuryl fluoride also, the egg stage of insects has been reported to be highly tolerant (Kenaga. 1957; Bell *et al.,* 2002). It has to be examined whether stored grain insects will develop resistance to sulfuryl fluoride quickly (following exposure to sub-lethal doses in 3-5 generations) as in the case of phosphine.

Acknowledgement

The authors thank the Director, Central Food Technological Research Institute, Mysore, for encouragement and support. The study was carried out during 2006-07 under the project "Substitutes for methyl bromide in pre-shipment fumigation of food commodities" sponsored by The Department of Agriculture and Cooperation, Ministry of Agriculture, Government of India. New Delhi.

References

Bell, C.H., 2006. Factors affecting the efficacy of sulfuryl fluoride as a fumigant. pp.519-526.In: Lorini I. et al (Eds.), Proceedings of The 9th International Working Conference on Stored Product Protection., 15-18 October, 2006, Campinas, Brazil, Brazilian Post-harvest Association, Campinas, Brazil,

Bell, C.H., Wontner-Smith, T.J. and Savvidou, N., 2002. Some properties of sulfuryl fluoride in relation to its use as a fumigant in the cereals industries. pp. 910-915 In:Credland et al., (Eds).Advances in Stored Product Protection, Proceedings of the 8th International Conference on Stored Product Protection, 22-26 July 2002, York, UK, CAB International, Wallingford, UK

Desmarchelier, J. M., 1994. Carbonyl sulfide as a fumigant for control of insects and mites. pp, 78-82.In Highley, E et al (Eds.),: Stored Product Protection, Proceedings of the 6th International Working Conference on Stored -product Protection, , 17-23 April, 1994, Canberra, Australia, CAB International, Wallingford, UK,

Guogan, X., Zhongmei, C, Zhao, S.and Nengzhi, Q., 1999. The development of sulfuryl fluoride (SO_2F_2) in China- a brief introduction pp. 562-566.In: Stored product Protection, Proceedings of the 7th International Working Conference on Stored Product Protection, 14-19 October 1998, Beijing, China, Sichuan Publishing House of Science & Technology, Chengdu, China,

Kenaga, E.E, 1957. Some biological, chemical and physical properties of sulfuryl fluoride as an insecticidal fumigant. *Econ. Entomol,*, 50:1-6.

MBTOC 2002. United Nations Environment Programme (UNEP) Methyl Bromide Technical Options Committee (MBTOC) 2002 Assessment of the Alternatives to Methyl bromide. UNEP, Nairobi, 435 pp.

Prabhakaran, S., 2006. Commercial performance and global development status of Profume® gas fumigant. pp.635-641In: Lorini, I. et al (Eds.), Proceedings of The 9th International Working Conference on Stored Product Protection., 15-18 October, 2006, Campinas, Brazil, Brazilian Post-harvest Association, Campinas, Brazil,.

Rajendran, S., 2001. Alternatives to methyl bromide as fumigants for stored food commodities. *Pesticide Outlook*, 12: 249-253.

Rajendran, S., Shrilatha, B., Rekha-Savanth,T.L. and Nayak, K., 2000. Response of field strains of stored grain insect pests to phosphine fumigant. *Pestic Res.,* 12:63-67.

Subramanyam, B and Hagstrum, D.W., 1996. Resistance measurement and management. pp. 331-397.In: Subramanyam, B., Hagstrum D.W. (Eds), Integrated Management of Insects in Stored Products, Marcel Dekker, Inc., New York.

— xxx —

24

Screening of Different Wheat Varieties for Susceptibility to Lesser Grain Borer, *Rhizopertha dominica* (Fab.) Coleoptera: Bostrichidae Under Storage Conditions

KHER, R. H. AND R. C. JHALA
Department of Entomology, B. A. College of Agriculture, Anand Agricultural University, Anand – 388 110

SUMMARY

Preference of wheat varieties by R. *dominica* was in the order of varieties for their susceptibility: Lok-1 (4.97) > GW-496 (4.08) > GW-1139 (2.92) > GW-120 (2.69) > GW-322 (2.66) > GW-190 = GW-273 (2.40) > GW-173 (2.22) > Raj-1555 (1.88) > HI-8498 (1.66). The varieties were categorized into tolerant, susceptible, more susceptible and highly susceptible. None of the tested wheat varieties was observed to be immune against *R. dominica*.

Key words: Oviposition Preference, Population Growth, Germination Loss.

1. INTRODUCTION

The information on relative susceptibility of wheat varieties to stored grain pests serve as guidance for suitable protective measures with special reference to highly susceptible varieties. In the last 15 years, numbers of wheat varieties were screened for their susceptibility to *Rhizopertha dominica* Fab. n India (Saxena and Singh, 1995; Saxena *et al*., 1996; Jha *et al*., 1999; Sharma *et al*., 2001; Singh *et al*., 2001, 2003 and 2005). However, research on identification of wheat varieties resistant or less susceptible to the stored-grain pests should be continued since new varieties/genotypes are released continuously over a period of time. Definite information on the relative susceptibility of the recent wheat varieties/genotypes to *R. dominica* is not available, hence the study.

2. MATERIALS AND METHODS

Ten varieties of wheat (Lok-1, GW-120, GW-173, GW-190, GW-273, GW-496, GW-322, GW-1139, Raj-1555 and HI-8498) were evaluated for their susceptibility to *Rhizopertha dominica* L. All the varieties were procured from Main Wheat Research Station, Sardarkrushinagar Dantiwada Agricultural University, Vijapur (District Mehsana), Gujarat. The wheat varieties were evaluated for their susceptibility to *R. dominica* based on oviposition preference, population growth and loss during storage as below.

2.1 Evaluation Based on Oviposition Preference

An experiment was carried out during January 2004 and repeated during April 2004 in Completely Randomized Design with four repetitions. In each replication, 200 bold grains *i.e*., 20 grains of each variety were selected randomly and fixed (using solution of Acacia gum) randomly on white paper sheet (10 x 20 cm) in such a way that each grain remains 1 cm apart from each other and crease on the top. For the purpose, 10 x 20 cm area of paper sheet was divided into 200 square blocks and each grain was fixed in the centre of the block. Such four sheets, one sheet for one repetition were prepared. Each sheet with grains was placed in galvanized cage (Diameter: 35 cm, Height: 10.5) individually. Ten pairs of *R. dominica* (5 –10 days old) obtained from laboratory culture were released in each cage for egg laying and cage was covered with two fold muslin cloth kept in position by means of rubber band to prevent the adults from escaping. After 5 days of oviposition, the adults were discarded from each cage and the observations on number of eggs laid on 20 grains were observed under binocular microscope. The data on number of eggs laid on 20 grains under each repetition of different varieties were subjected to ANOVA after transforming them to square root. The data of two separate experiments were also pooled.

2.2. Evaluation based on population growth

The experiment was carried during April – September, 2004 in Completely Randomized Design (CRD) with 3 replications. For each variety, three samples of grains each of 50 g (one sample for one repetition) were filled in plastic tube (Diameter: 7.00 cm and Height: 8.00 cm) individually. Ten pairs of *R. dominica* (5 to 10 days old) were released in each tube for egg laying and each tube was covered with two fold muslin cloth kept in position by means of rubber band to prevent the adults from escaping. After 7 days of oviposition, the adults were discarded from each tube. The observations on number of adults (live + dead) developed in each repetition were made after 5 month of storage and the data were subjected to ANOVA after transforming them to square root.

2.3. Evaluation based on weight loss

For the purpose, 100 grains were selected randomly from each sample and segregated into weevilled grain and germ eaten grain. The weevilled, germ eaten grain and 100 undamaged grains were weighed and the per cent loss in weight was calculated sample/ repelications wise using the formula: $L = (W + G) - 100/S\ (W1 + G1)$, where L = Per cent loss in weight, W = Percentage (by number) of weevilled grain, G = Percentage (by number) of germ eaten grain, S = Weight of 100 undamaged grain, W1= Weight of weevilled grains, G1= Weight of germ eaten grains (Srivastava *et al.*, 1973). The data on per cent loss in weight were subjected to ANOVA after transforming them to arc sin.

2.4. Evaluation based on germination loss

Germination tests were carried out before and after 5 months of storage. In each test, three samples (one sample as one replication), each of 100 seeds of wheat grains selected randomly were drawn from each variety. The test was carried out in circular Whatman No. 1 filter paper kept in petri plates (Diameter 10 cm) and moist for germination of seeds. The seeds were spread on the germination paper at uniform distance in the petri-plate. The petri-plate was covered with the lid carrying moist filter paper and kept in a seed germinator maintained at 21 ± 1° C temperatures and 95 ± 2 % relative humidity. Little quantity of distilled water was sprinkled on filter paper once a day to keep it moist. The numbers of grains germinated were counted after 7 days of incubation. The per cent loss in germination was worked out repetition wise for each variety based on the data on germination before and after 5 months of storage. The data were subjected to ANOVA after transforming them to arc sin.

2.5. Categorization of varieties

The wheat varieties were grouped into four categories of susceptibility to *R. dominica viz.*, tolerant, susceptible, more susceptible and highly susceptible based on four parameters *viz.*, oviposition preference (number of eggs per 20 seeds), population growth (number of adults emerged during 5 months of storage), per cent loss in weight and per cent loss in germination. For the purpose mean value of individual variety (i) was compared with mean value of all varieties () and standard deviation (SD) following the scale as adopted by Patel *et al.* (2002). The transformed data were used for computation of i, and SD in case of each parameter. The scale used for categorizing different varieties was as under.

Category of resistance	**Scale for resistance**
Tolerant	$\bar{X}_i < \bar{X}$
Susceptible	$\bar{X}_i > \bar{X} < (\bar{X} + 1SD)$
More susceptible	$\bar{X}_i > (\bar{X} + 1SD) < (\bar{X} + 2SD)$
Highly susceptible	$\bar{X}_i < (\bar{X} + 1SD)$

3. RESULTS AND DISCUSSION

3.1. Evaluation based on oviposition preference

The data on number of eggs/20 seeds for individual experiments as well as pooled are presented in Table 1. The pooled data revealed that there was significant difference among varieties. The chronological order of their susceptibility with average number of eggs/20 seeds (given in bracket after each variety) was Lok-1 (4.97) > GW-496(4.08) > GW-1139(2.92) > GW-120(2.69) > GW-322(2.66) > GW-190 = GW-273(2.40) > GW-173(2.22) > Raj-1555(1.88) > HI-8498(1.66). GW-496 was at par with Lok-1 on one side while with GW-1139 on another side of chronological order. Similarly, GW-190, GW-273 and GW-173 were at par with each other as well as with GW-322, GW-120 and GW-1139 on one side while with Raj-1555 and HI-8498 on another side of chronological order of varieties for their susceptibility to *R. dominica*. The performance of varieties was consistent in two separate experiments.

Table 1 : Susceptibility of wheat varieties to *R. dominica* based on oviposition preference

Varieties	**Number of eggs laid on 20 seeds**		
	Experiment-1	**Experiment-2**	**Mean**
Lok-1	2.49 a (6.20)	1.98 a (3.92)	2.23 a (4.97)
GW-496	2.34 a (5.47)	1.70 ab (2.89)	2.02 ab (4.08)
HI-8498	1.49 b (2.22)	1.10 c (1.21)	1.29 d (1.66)
Raj-1555	1.49 b (2.22)	1.26 bc (1.59)	1.37 d (1.88)
GW-120	1.72 b (2.95)	1.56 abc (2.43)	1.64 c (2.69)
GW-273	1.63 b (2.65)	1.48 bc (2.19)	1.55 cd (2.40)
GW-1139	1.79 b (3.20)	1.63 ab (2.66)	1.71 bc (2.92)
GW-190	1.70 b (2.89)	1.40 bc (1.96)	1.55 cd (2.40)
GW-173	1.70 b (2.89)	1.27 bc (1.61)	1.49 cd (2.22)
GW-322	1.70 b (2.89)	1.56 abc (2.43)	1.63 c (2.66)
Mean	1.80 (3.24)	1.49 (2.22)	1.65 (2.72)
S. Em. *± Variety (V)*	0.15	0.17	0.11
Experiment (E)	-	-	0.05
V x E	-	-	0.16
CD at *5 % V*	0.43	0.49	0.32
E	-	-	0.14
V x E	-	-	NS
CV %	16.50	22.59	19.30

Means in parentheses are retransformed values, those outside are transformed values.
Means with letter(s) in common do not differ significantly in respective column.
NS: Not significant.

3.2. Evaluation based on population growth

The data on number of adults developed due to initial oviposition in a week by 10 pairs of *R. dominica* and after 5 months of wheat storage in laboratory are presented in Table 2. There was significant difference among varieties. The chronological order of varieties for their susceptibility to *R. dominica* based on number of adults developed (given in bracket after each variety) after 5 months of storage period was : GW-496(108.78) > Lok-1(101.40) > HI-8498(100.00) > GW-1139(69.89) > GW-120(62.73) > GW-173(47.75) > GW-273(44.09) >Raj-1555(42.51) > GW-322(27.35) > GW-190(26.62). The varieties, Lok-1,

HI-8498 and GW-1139 were at par with each other as well as with GW-496 on one side while with GW-120 on another side of chronological order. Similarly, the varieties, GW-173, GW-273 and Raj-1555 were at par with each other as well as with GW-120 and GW-1139 on one side while with GW-322 and GW-190 on another side of chronological order of varieties based on population growth.

Table 2 : Susceptibility of wheat varieties to *R. dominica* based on population growth and weight loss after 5 months of storage

Varieties	**Number of adults emerged***	**Per cent Weight Loss****
1	**2**	**3**
Lok-1	10.07 ab (101.40)	26.72 a (20.22)
GW-496	10.43 a (108.78)	25.75 a (18.87)
HI-8498	10.05 ab (100.00)	24.70 ab (17.46)
Raj-1555	6.52 cd (42.51)	23.26 bc (15.59)
GW-120	7.92 bc (62.73)	22.56 bcd (14.72)
GW-273	6.64 cd (44.09)	21.39 cde (13.30)
GW-1139	8.36 abc (69.89)	21.35 cde (13.25)
GW-190	5.16 d (26.62)	21.16 cde (13.03)
GW-173	6.91 cd (47.75)	20.36 de (12.10)
GW-322	5.23 d (27.35)	19.14 e (10.75)
S. Em. ±	0.76	0.81
CD 5%	2.23	2.40
CV %	16.93	6.22

Means in parentheses are retransformed values, those outside are * and arc sin** transformed values.
Means with letter(s) in common do not differ significantly in respective column.

3.3. Evaluation based on weight loss

The data on per cent loss in weight due to infestation by *R. dominica* in wheat grain after 5 month of storage period presented Table 2. There was significant difference among varieties. The chronological order of varieties for their susceptibility to *R. dominica* based on per cent weight loss was: Lok-1(20.22 %) > GW-496(18.87 %) > HI-8498 (17.46 %) > Raj-1555(15.59 %) > GW-120(14.72 %) > GW-273 13.30 %) > GW-1139(13.25 %) > GW-190(13.03 %) > GW-173(12.10 %) > GW-322(10.75 %). The variety HI-4898 was at par with GW-496 and Lok-1 on one side and with Raj-1555 and GW-120 on another side of chronological order. The later two varieties were at par with GW-273, GW-1139 and GW-190. The varieties GW-273, GW-1139, GW-190 and GW-173 were at par with each other as well as with GW-120 on one side and with GW-322 on another side of chronological order of varieties based on weight loss.

3.4. Evaluation Based on Germination Loss

The data on per cent germination in different varieties of wheat before artificial infestation, after 5 months of storage under infestation and per cent loss in germination due to infestation by *R. dominica* are presented in Table 3. The data on per cent germination before artificial infestation revealed that the per cent germination among varieties ranged from 95.00 to 95.65 with an average of 95.33. There was non-significant difference among the varieties indicating uniform germination among the varieties. After 5 month of artificial infestation of *R. dominica*, there was significant adverse effect on germination of different varieties of wheat. The chronological order of varieties for their susceptibility to *R. dominica* based on per cent germination was: Lok-1 (25.27 %) > GW-496(38.62 %) > GW-273(40.96 %) > Raj-1555(41.28 %) > GW-322(41.30 %) > GW-120(43.23 %) > GW-190(47.94 %) > GW-1139(50.64 %) > HI-8498(53.70 %) > GW-173(60.04 %). Lok-1 recording lowest per cent germination was highly susceptible but it was at par with GW-496, which was in turn at par with rest of the varieties except GW-173. GW-173 recording highest per cent germination was found least susceptible but, it was at par with HI-8498, GW-1139 and GW-190.

The data on per cent loss in germination after 5 months of storage of varieties under artificial infestation are presented Table 3. The ANOVA clearly revealed that there was significant difference among the varieties. The chronological order of varieties for their susceptibility to *R. dominica* based on per cent loss in germination was: Lok-1 (70.25 %) > GW-496 (56.97 %) > GW-322 (54.30 %) > GW-273 (53.97 %) > GW-120(52.32 %) > GW-190 (47.30 %) > Raj-1555 (47.26 %) > GW-1139 (43.56 %) > HI-8498 (41.14 %) > GW-173 (35.25 %). Lok-1 recording highest loss in per cent germination was found most susceptible and it was at par with GW-496, GW-322 and GW-273. The varieties, GW-190, Raj-1555, GW-1139 and HI-8498 were at par with each other as well on as with GW-120, GW-273; GW-322 and GW-496 on one side while with GW-173 on another side of chronological order of varieties for their susceptibility to *R. dominica*.

Table 3 : Susceptibility of wheat varieties to *R. dominica* based on per cent loss in germination

Treatments	Per cent germination		Per cent loss in germination
	Before infestation	After 5months of storage	
Lok-1	77.97 (95.65)	30.18 c (25.27)	56.95 a (70.25)
GW-496	77.97 (95.65)	38.42 bc (38.62)	49.01 ab (56.97)
HI-8498	77.09 (95.00)	47.12 ab (53.70)	39.90 bc (41.14)
Raj-1555	77.61 (95.40)	39.98 b (41.28)	43.43 bc (47.26)
GW-120	77.97 (95.65)	41.11 b (43.23)	46.33 b (52.32)
GW-273	77.09 (95.00)	39.79 b (40.96)	47.28 ab (53.97)
GW-1139	76.21 (94.32)	45.37 ab (50.64)	41.30 bc (43.56)
GW-190	77.50 (95.31)	43.82 ab (47.94)	43.45 bc (47.30)
GW-173	77.97 (95.65)	50.79 a (60.04)	36.42 c (35.25)
GW-322	77.97 (95.65)	39.99 b (41.30)	47.47 ab (54.30)
Mean	77.53 (95.33)	41.66 (44.30)	45.15 (50.23)
S. Em. ±	0.64	3.20	3.30
CD 5 %	NS	9.45	9.74
CV %	1.44	13.31	12.67

Means in parentheses are retransformed values, those outside are arc sin transformed values. Means with letter(s) in common do not differ significantly in respective column.
NS: Not significant.

3.5. Categorization of Varieties for Susceptibility

The details of categorization of different verities of wheat are presented in Table 4. Based on oviposition preference (number of eggs/

Table 4 : Categorization of different varieties of wheat for their susceptibility to *R. dominica* based on different parameters

Based on number eggs per 20 seeds :	$\bar{X}$ **= 2.79**	**SD = 1.00**
Tolerant	$\bar{X}_i$ < 2.79	HI-8498(1.66), Raj-1555(1.88), GW-173(2.22), GW-273(2.40), GW-190(2.40), GW-322(2.66) and GW-120(2.69)
Susceptible	$\bar{X}_i$ > 2.79 < 3.79	GW-1139(2.92)
More susceptible	$\bar{X}_i$ > 3.79 < 4.79	GW-496(4.08)
Highly susceptible	$\bar{X}_i$ > 4.79	Lok-1(4.97)
Based on number of adults emerged:	$\bar{X}_i$ **= 63.11**	**SD = 30.90**
Tolerant	$\bar{X}_i$ **< 63.11**	GW-190(26.62), GW-322(27.35), Raj-1555(42.51), GW-273(44.09), GW-173 (47.75) and GW-120 (62.73)
Susceptible	$\bar{X}_i$ **> 63.11 < 94.01**	GW-1139(69.89)
More susceptible	$\bar{X}_i$ **> 94.01 < 124.91**	HI-8498(100.00), Lok-1(101.40) and GW-496(108.78)
Highly susceptible	$\bar{X}_i$ > 124.91	–
Based on per cent loss in weight:	$\bar{X}_i$ **= 14.93**	**SD = 3.07**
Tolerant	$\bar{X}_i$ **< 14.93**	GW-322(10.75), GW-173(12.10), GW-190(13.03), GW-1139(13.25), GW-273(13.30) and GW-120(14.72)
Susceptible	$\bar{X}_i$ **> 14.93 < 18.00**	Raj-1555 (15.59) and HI-8498 (17.46)
More susceptible	$\bar{X}_i$ **> 18.00 < 21.07**	GW-496(18.87) and Lok-1(20.22)
Highly susceptible	$\bar{X}_i$ **> 21.07**	-
Based on per cent loss in germination:	$\bar{X}_i$ **= 50.23**	**SD = 9.72**
Tolerant	$\bar{X}_i$ < 50.23	GW-173(35.25), HI-8498(41.14), GW-1139 (43.56), Raj-1555(47.26) and GW-190(47.30)
Susceptible	$\bar{X}_i$ > 50.23 < 59.95	GW-120(52.32), GW-273(53.97), GW-322(54.30) and GW-496(56.97)
More susceptible	$\bar{X}_i$ > 59.95 < 69.67	-
Highly susceptible	$\bar{X}_i$ > 69.67	Lok-1(70.25)

20 seeds of wheat), HI-8498, Raj-1555, GW-173, GW-273, GW-190, GW-322 and GW-120 recording less than 2.79 eggs/20 seeds were found tolerant varieties. While the varieties, GW-1139 recording less than 3.79 but more than 2.79 eggs/20 seeds; GW-496 recording less than 4.79 but more than 3.79 eggs/20 seeds and Lok-1 recording more than 4.79 eggs/20 seeds were found less, more and highly susceptible varieties, respectively.

Based on population growth (number of adults emerged) during 5 months of storage period, GW-190, GW-322, Raj-1555, GW-273, GW-173 and GW-120 recording less than 63.11 adults were found tolerant varieties. The variety GW-1139 recording more than 63.11 but less than 94.01 adults were found susceptible, while HI-8498, Lok-1 and GW-496 recording more than 94.01 but less than 124.91 adults were found more susceptible varieties.

Based on per cent loss in weight, GW-322, GW-173, GW-190, GW-1139, GW-273 and GW-120 recording less than 14.93 % loss were found tolerant varieties. The varieties Raj-1555 and HI-8498 recording more than 14.93 but less than 18.00 % loss were found susceptible, while GW-496 and Lok-1 recording more than 18.00 but less than 21.07 % loss were more susceptible to *R. dominica.*

The per cent losses in germination indicate indirectly the preference of *R. dominica* for feeding on germ/embryo. The varieties recording per cent loss in germination below 50.23 were: GW-173, HI-8498, GW-1139, Raj-1555 and GW-190 were found tolerant, while varieties GW-120, GW-273, GW-322 and GW-496 recording per cent loss in germination above 50.23 but below 59.95 were susceptible. The variety Lok-1 recording more than 69.67 % loss in germination was found highly susceptible to *R. dominica.*

Various workers carried out studies on relative resistance/susceptibility of wheat varieties against *R. dominica* in India since last 25 years Zaz *et al.*, 1982; Saxena and Singh, 1995; Saxena *et al.*, 1996; Jha *et al.*, 1999; Sharma *et al.*, 2001; Singh *et al.*, 2001 and Singh *et al.*, 2005). However, none of them could find a variety completely resistant to the insect attack. The present results on relative susceptibility of wheat varieties to *R. dominica* could not be compared with the results of work done elsewhere because of uncommonness in varieties. However, based on number of adults emerged and per cent loss in weight, Lok-1 was found more susceptible to *R. dominica* (Patel, 1983). Similarly, based on number of adults emerged, per cent loss in weight and per cent germination of infested seeds, the order of susceptibility to *R. dominica*, of 3 varieties of wheat was HI- 8498 < GW-173 < Lok-1 as reported by Sharma *et al.* (2001). Raj-1555 was found least susceptible to *R. dominica* as per report of Singh *et al.* (2005). These reports on susceptibility of wheat varieties tally with present results *i.e.*, higher susceptibility of Lok-1 and tolerant of GW-173 and Raj-1555 to *R. dominica.*

The resistance/tolerance may be due to either fewer eggs laid initially or elimination of immature stages during developmental period to adult emergence as reported by Phadke and Bhatia (1974). Variation in developmental period might be due to antibiosis factor. It is indicated that probably some physical characteristics of the grains such as hardness which prevent the larvae from boring inside make the variety tolerant. Lower pelshenke value (grain hardness) indicating low strength of gluten could be associated with susceptibility (Bains, 1971 and Patel, 1983). Developmental period is reported faster in varieties with higher protein content and higher moisture (Khokhar and Gupta, 1974 and Patel, 1983). No significant relationship between seed size and adult emergence could be established by Sharma *et al.*, (2001). It is possible that both phenomena *i.e.*, non-preference and antibiosis might be present simultaneously which impart tolerance or low susceptibility to wheat varieties against *R. dominica.*

Overall, it can be concluded that the varieties GW-173 and GW-190 by all the four parameters can be considered as tolerant to *R. dominica*. The variety Lok-1 was found highly susceptible to *R. dominica*. None of the tested wheat varieties was observed to be immune against *R. dominica.*

References

Bains, S.S. 1971. Effect of temperature and moisture on the biology of *Rhizopertha dominica* Fab. (Bostrichidae: Coleoptera). *Bull. Grain Tech.*, 9(4): 257-264.

Jha, A.N., S.C. Khanna and S. Singh 1999. Olfactory response of insect pests of stored grains to wheat cultivars. *Indian J. Ent.*, 61(3): 288-290.

Khokhar, D.S. and D.S. Gupta 1974. Relative resistance of some varieties of wheat to *Sitophilus oryzae* (L.) and *Rhizopertha dominica* F. at different temperatures. *Bull. Grain Tech.*, 12(2): 117-123.

Patel, I.S., B.G. Prajapati, G.M. Patel and A.R. Pathak 2002. Response of castor genotypes to castor semilooper, *Achaea janata* Fab., *J. Oilseeds Res.*, 19 (1): 153.

Patel, M.A. 1983. The assessment of the inherent vulnerability of wheat varieties to post harvest infestation by *Trogoderma granarium* E. and *Rhizopertha dominica* F. under various ecological conditions. *Ph. D. (Agri.) Thesis*, GAU, Anand.

Phadke, K.G. and S.K. Bhatia 1974. Population growth of *Sitophilus oryzae* (L.) and *Rhizopertha dominica* F. in wheat varieties. *Indian J. Ent.*, 36(4): 251-260.

Saxena, A. and Y.P. Singh 1995. Correlation studies of losses in wheat varieties and reproduction potential of *Rhyzopertha dominica* Fabr. *Advances in Agricultural Research in India*, 3: 42-50.

Saxena, A., Y.P. Singh and N.D. Pandey 1996. Studies on growth and development of *Rhizopertha dominica* on wheat varieties. *Bull. Ent.* (New Delhi), 37(1-2): |72-77.

Sharma, V., N.S. Bhadauria and S.S. Jhakmola 2001. Reaction of some wheat varieties to lesser grain borer, *Rhizopertha dominica*. *Indian J. Ent.*, 63(2): 163-165.

Singh, A.K., C. Singh and V. Pandey 2001. Weight loss in grains of different wheat varieties due to infestation of *Rhizopertha dominica* Fabricius (Coleoptera: Bostrichidae). *Shashpa*, 8(1): 81-82.

Singh, B., D. Singh, Ashok kumar and Dhirender. 2003. Extent of damage in stored wheat varieties caused by *Rhizopertha dominica (*Fab.*)* at different time interval. *Current Agriculture*, 27(1/2): 111-113.

Singh, R.K., A.K. Singh and H.K. Singh 2005. Studies on protein composition of different wheat varieties for the preference of *Rhizopertha dominica* (Feb.) *Biochemical and Cellular Archives*, 5(2): 207-212.

Zaz, G.M., S.C. Bhardwaj and C.P.S. Yadava 1982. Relative susceptibility of some wheat varieties to the lesser grain borer, *Rhizopertha dominica* Fabricius. *Indian J. Ent.*, 44(1): 77-82.

— xxx —

25

Eco-friendly Approaches for the Management of Rice-Weevil, *Sitophilus oryzae* (L.) in Pop Sorghum Stored in Jute Bag

KAVITA JADHAV[1] AND SHEKHARAPPA[2]
[1] Research Scholar [2] Professor
AICSIP, Sorghum Research Scheme, University of Agricultural Sciences
Dharwad-580005, Karnataka

SUMMARY

The study clearly indicated that sweet flag powder (1%) was superior as it showed maximum adult mortality, least per cent grain damage and weight loss. The next best treatments were neem seed powder @ 5% and custard seed powder @ 5% recording 86.6, and 83.3 per cent adult mortality, 8.11 and 6.50 per cent grain damage and 2.2 and 2.1 per cent weight loss, respectively. Among different inert materials, kaolinite (10%) was found relatively good but inferior to the above plant products.

Key words: Grain Damage, Kaolinite Clay, Grain Protectants

1. INTRODUCTION

The pop sorghum grains in storage are attacked and damaged by a number of pests including *Sitophilus oryzae* (L.) a number one pest. The adults and larvae are internal feeders and cause great loss to the grain, both in quality and quantity. It causes enormous losses up to 100 per cent to stored grains (Singh *et al.*, 1975). This pest not only causes untold damage by its own activity but in doing so also provides food and conditions that are suitable to the introduction and development of other grain infesting insects. Since, the pop sorghum grains are not consumed immediately as they are stored for long time and sold as snack food throughout the year. In the present investigation, certain grain protectants were evaluated for their efficacy against *S. oryzae*.

2. MATERIAL AND METHODS

An experiment was conducted to study the effect of different grain protectants on seed storability of pop sorghum variety Talakal-6. The experiment was initiated with 17 treatments replicated thrice. The treatments comprised of products of 10 plant species *viz.* pongam leaf powder (5%) (*Pongamia pinnata* (L.) Pierre), adhatoda leaf powder (5%) (*Adhatoda vasica* Medi.), periwinkle leaf powder (5%) (*Catharanthus roseus* (L.) Don.), neem leaf powder (5%) (*Azadirachta indica* An.dr. Juss.,) Indian Privet leaf powder (5%) (*Vitex negundo* L.), sweet flag rhizome powder (1%) (*Acorus calamus* L.), neem seed powder (5%), tulsi leaf powder (5%) (*Ocimum sanctum* (L.)), custard apple seed powder (5%) (*Annona squmosa* (L.)), turmeric (5%) (*Curcuma domestica* Valeton.), kaolinite (10%) and five inert materials, ash (30%), bentonite (10%), saw dust (10%), sand (10%), malathion (5%) and an untreated check. Small jute bags that could hold 250 g of pop sorghum grains were prepared. 100 g of grains of pop sorghum variety Talakal-6 were treated with the plant products and inert materials at the mentioned dosage and taken in the jute bags, 10 pair of adult weevils were released and bags were tied. Samples were drawn at monthly intervals for recording observation on per cent mortality, per cent weight loss, per cent grain damage, germination and population buildup at 30, 60, 90, 120, 150 and 180 days after storage (DAS).

3. RESULTS AND DISCUSSION

Among all the grain protectants, sweet flag at 1% afforded maximum protection over other protectants. It was found highly effective against *S oryzae* causing cent percent mortality at 180 days after storage. The other best treatments were neem seed powder and custard seed powder at 5% which recorded cent per cent mortality upto 90 days after storage and later on declined to 86.6 and 83.3 per cent, respectively at 180 days after storage (Table1). No adult emergence, seed damage and and weight loss was recorded even at 180 days after storage in gunny bags when treated with sweet flag. (Tables-2, 3 & 4). Similarly Sivasrinivasu (2001) reported 100 per cent

Table 1 : Evaluation of grain protectants against *S. oryzae* at various pe riods of storage in jute bag

Treatments	Per cent adult mortality at						Mean
	30 DAS	60 DAS	90 DAS	120 DAS	150 DAS	180 DAS	
Pongam leaf powder	46.6f (43.05)	50.0e (45.00)	61.6e (51.71)	48.3h (44.03)	40.0g (39.23)	38.3g (38.23)	47.47f
Adhatoda leaf powder	41.6g (40.16)	45.0f (42.13)	50.0g (45.00)	38.3i (38.23)	28.3h (32.14)	21.6i (27.69)	37.46h
Catharanthus leaf powder	43.3g (41.15)	43.3f (41.15)	56.6f (48.79)	46.6h (43.05)	43.3f (41.15)	26.6h (31.05)	43.28g
Neem leaf powder	63.3e (52.12)	76.6d (61.07)	76.6d (60.40)	60.0f (50.77)	50.0e (45.00)	41.6fg (40.16)	61.35e
Indian Privet leaf powder	65.0d (53.73)	80.0c (63.44)	80.0c (60.40)	53.3g (46.89)	48.3e (44.03)	46.0f (42.71)	62.10e
Sweet flag	100.0a (90.0)	100a (90.0)	100.0a (0.00)	100a (0.00)	100a (90.00)	100a (90.00)	100.0a
Neem seed powder	100.0a (90.0)	100a (90.0)	100a (90.0)	91.6c (73.15)	90.0b (71.56)	86.6b (68.53)	94.70b
Tulsi leaf powder	16.6h (24.04)	20.0h (26.56)	33.3h (35.24)	25.0k (30.00)	21.6j (27.69)	18.3ij (25.33)	22.46i
Custard seed powder	100.0a (90.0)	100a (90.0)	100a (90.0)	91.6c (73.15)	90.0b (71.56)	83.3c (65.88)	94.15b
Turmeric	13.3i (21.39)	16.6i (24.04)	20.0j (26.56)	33.3j (75.00)	21.6j (27.69)	21.6i (27.69)	21.06i
Kaolinite	92.3b (73.89)	90.0b (71.56)	85.6b (67.78)	80.0d (63.44)	73.3c (58.89)	75.0d (60.00)	82.70c
Bentonite	85.6c (67.70)	90.0b (71.56)	85.0b (67.21)	73.3e (58.89)	70.0d (56.79)	63.3e (52.71)	77.86d
Ash (cow dung)	17.3h (23.21)	26.6g (31.05)	30.0i (33.21)	21.6l (27.69)	20.0j (26.56)	16.6j (24.04)	22.01i
Sand	13.3i (21.39)	16.6i (24.09)	20.0j (26.56)	30.0k (33.21)	25.0i (30.00)	18.3ij (25.33)	20.53i
Saw dust	10.0j (18.44)	0.0j (0.00)	0.0k (0.00)	0.0m (0.00)	0.0k (0.00)	0.0k (0.00)	1.66j
Malathion	100.0a (90.0)	100.0a (90.0)	100a (0.00)	93.3b (75.00	88.3b (70.00)	81.6c (64.60)	93.86b
Untreated control	0.0k (0.00)	0.0j (0.00)	0.0k (00.0)	0.0m (0.00)	0.0k (0.00)	0.0k (0.00)	00.00j
CD (1%)	1.54	1.66	1.55	1.65	1.67	2.56	3.87
SEm±	0.37	0.43	0.40	0.43	0.43	0.66	1.00

DAS - Days after storage

Means followed by the same letter do not differ significantly by DMRT (P=0.01) Figures in the parentheses are arc sine transformed value

Table 2 : Efficacy of grain protectants against *S. oryzae* at different periods of storage in jute bag

Treatments	Per cent gram damage						Mean
	30 DAS	60 DAS	90 DAS	120 DAS	150 DAS	180 DAS	
Pongam leaf powder	15.3f (23.03)	17.8b (24.95)	20.3e (26.78)	25.0b (30.00)	38.6f (38.41)	50.0e (45.00)	27.84d
Adhatoda leaf powder	16.6de (24.04)	19.2cd (25.99)	21.8f (24.83)	29.0a (35.58)	44.33e (41.73)	53.3d (46.89)	30.70c
Catharanthus leaf powder	22.6b (28.38)	24.3b (29.53)	26.4ab (30.92)	34.6a (36.03)	50.0cd (45.00)	62.3c (52.12)	36.70b
Neem leaf powder	15.3f (23.03)	17.6b (24.80)	21.0e (27.28)	23.3b (28.79)	34.3h (35.85)	42.0f (40.40)	25.63e
Indian Privet leaf powder	16.0ef (23.58)	18.4d (25.40)	21.3de (27.49)	23.3b (28.79)	36.3g (37.05)	45.0g (42.13)	26.71de
Sweet flag	0.0l (0.00)	0.0j (0.00)	0.0k (0.00)	0.0f (0.00)	0.0m (0.00)	0.0k (0.00)	0.00j
Neem seed powder	1.7j (7.43)	2.43h (8.91)	3.4i (10.63)	4.0de (11.54)	6.1k (14.30)	8.11j (6.59)	4.28hi
Tulsi leaf powder	17.5cd (24.73)	20.7c (27.06)	26.1ab (30.72)	39.6a (398.0)	54.60b (47.04)	63.3c (52.71)	36.96b
Custard seed powder	1.0k (5.74)	1.6i (7.27)	2.3j (8.72)	36.00e (9.98)	4.0l (11.54)	6.50j (15.89)	3.06i
Turmeric	21.3b (27.44)	23.4b (29.60)	25.2bc (30.13)	33.3a (35.37)	51.30c (45.75)	65.5b (54.09)	36.68b
Kaolinite	3.0i (9.98)	4.4g (12.11)	5.0h (12.92)	6.1cd (14.30)	8.0j (16.43)	10.0i (18.44)	6.08g
Bentonite	5.5h (13.56)	6.1f (14.3)	6.9g (15.23)	8.0c (16.43)	11.90i (20.18)	13.0h (21.13)	8.56f
Ash (cow dung)	13.2g (21.30)	15.1e (22.87)	17.5f (24.73)	20.7b (27.06)	53.3b (46.89)	67.0b (54.94)	31.13c
Sand	14.8f (22.63)	17.3d (24.58)	19.6ef (26.28)	22.6a (28.38)	37.30fg (37.68)	51.3e (45.75)	26.16e
Saw dust	17.9c (25.63)	20.6c (26.99)	23.3cd (28.86)	29.0a (35.58)	49.30d (44.60)	54.60b (47.06)	30.76c
Malathion	1.9j (7.92)	2.7h (9.46)	4.3hi (11.97)	5.50cd (13.56)	6.1k (14.30)	8.0j (16.43)	4.75h
Untreated control	25.4a (30.33)	26.8a (31.18)	28.4a (32.20)	37.3a (37.68)	64.0a (23.13)	73.00a (58.69)	42.48a
CD (1%)	0.94	1.39	1.42	2.96	1.08	1.13	1.22
SEm±	0.24	0.36	0.37	0.77	0.28	0.29	0.32

DAS – Days after storage

Means followed by the same letter do not differ significantly by DMRT (P=0.01) Figures in the parentheses are arc sine transformed value

Table 3 : Per cent weight loss of Pop sorghum due to *S. oryzae* infestation in jutebag treated with grain protectants

Treatments	Per cent weight loss of pop sorghum at						Mean
	30 DAS	60 DAS	90 DAS	120 DAS	150 DAS	180 DAS	
Pongam leaf powder	4.5 f (12.25)	5.4e (13.44)	7.3de (15.68)	9.6de (18.05)	11.2gh (19.55)	12.0g (20.27)	8.33ef
Adhatoda leaf powder	5.5e (13.56)	6.8d (15.12)	7.76de (16.11)	10.3d (18.72)	12.1fg (20.36)	13.0g (21.13)	9.24de
Catharanthus leaf powder	4.6f (12.39)	7.6d (16.00)	8.26b (16.64)	9.2def (17.66)	13.9e (21.89)	15.2ef (22.95)	9.79d
Neem leaf powder	2.5g (9.10)	4.11f (11.68)	6.7e (15.00)	8.4ef (16.85)	13.1ef (21.22)	16.0e (23.58)	8.46ef
Indian Privet leaf powder	3.0g (9.98)	4.13f (11.68)	7.1 de (15.45)	8.3f (16.74)	12.0fg (20.27)	14.4f (22.30)	8.15f
Sweet flag	0.0j (0.00)	0.0i (0.00)	0.0h (0.00)	0.0j (0.00)	0.0m (0.00)	0.01 (0.00)	0.00i
Neem seed powder	0.0j (0.00)	0.0i (0.00)	0.0h (0.00)	0.9i (5.44)	1.8k (7.71)	2.2k (8.53)	0.81hi
Tulsi leaf powder	8.2cd (16.64)	10.2c (18.63)	12.5c (20.70)	14.9c (22.71)	18.0cd (25.10)	20.0c (26.56)	13.96c
Custard seed powder	0.0j (0.00)	0.0i (0.00)	0.0h (0.00)	0.0j (0.00)	1.101 (6.02)	2.1k (8.33)	0.53hi
Turmeric	7.6b (16.00)	12.6b (20.79)	16.6a (24.04)	18.3b (25.33)	20.0b (26.56)	22.0b (27.97)	16.18b
Kaolinite	1.3 i (6.55)	1.5h (7.04)	2.63g (8.13)	3.8g (11.24)	5.3i (13.31)	6.4i (14.65)	3.38g
Bentonite	1.8h (7.71)	2.13g (8.33)	2.8f (9.63)	4.5g (12.25)	6.0i (14.18)	8.2h (16.64)	4.23g
Ash (cow dung)	9.1bc (17.76)	10.2c (18.63)	12.0c (20.27)	14.2c (22.04)	14.6d (23.89)	18.3d (25.33)	13.36c
Sand	5.03ef (12.92)	5.6e (13.69)	7.5de (15.89)	8.7ef (17.16)	10.2h (18.63)	12.3g (20.53)	8.22ef
Saw dust	10.1b (18.53)	12.8b (20.96)	15.0b (22.79)	17.8b (24.95)	19.5bc (26.21)	20.3c (26.78)	15.92b
Malathion	0.0j (0.00)	0.0i (0.00)	1.0h (0.00)	1.8h (7.71)	2.5i (9.10)	3.2j (10.31)	1.25h
Untreated control	11.6a (19.91)	14.6a (22.46)	18.1a (25.10)	21.1a (27.35)	22.7a (28.45)	24.4a (29.60)	18.75e
CD (1%)	1.13	0.98	1.16	1.18	1.34	0.95	0.96
SEm±	0.29	0.25	0.30	0.31	0.35	0.24	0.24

DAS – Days after storage   Means followed by the same letter do not differ significantly by DMRT (P=0.01)   Figures in the parentheses are arc sine transformed value

Table 4 : Population buildup of adults of *S. oryzae* in jutebag treated with grain protectants

Treatments	Population buildup at						Mean
	30 DAS	60 DAS	90 DAS	120 DAS	150 DAS	180 DAS	
Pongam leaf powder	4.3i (2.19)	11.3e (3.43)	15.6cde (4.01)	26.6g (5.21)	38.0d (6.20)	54.0d (7.38)	24.96f
Adhatoda leaf powder	5.0h (2.34)	8.0h (2.91)	13.6de (3.75)	21.6h (4.70)	40.6d (6.41)	51.3de (7.19)	23.35f
Catharanthus leaf powder	7.0f (2.73)	11.3e (3.43)	17.0c (4.18)	33.6e (5.83)	53.0c (7.31)	64.3c (8.04)	31.03e
Neem leaf powder	4.3i (2.19)	8.3h (2.96)	15.0cde (3.93)	28.6f (5.40)	39.6d (6.33)	50.0de (7.10)	24.31f
Indian Privet leaf powder	6.0g (2.54)	9.0g (3.08)	13.3e (3.71)	22.0h (4.74)	37.0d (6.12)	45.0e (6.74)	22.05f
Sweet flag	0.0k (0.70)	0.0k (0.70)	0.0h (0.70)	0.0l (0.70)	0.0j (0.70)	0.0j (0.70)	0.00j
Neem seed powder	0.0k (0.70)	0.0k (0.70)	0.0h (0.70)	3.0k (1.87)	8.0h (2.91)	13.3h (3.71)	4.05i
Tulsi leaf powder	10.0d (3.24)	14.0b (3.80)	22.6b (4.80)	38.6d (6.25)	59.6b (7.75)	68.0c (8.27)	35.47cd
Custard seed powder	0.0k (0.70)	0.0k (0.70)	0.0h (0.70)	0.0l (0.70)	3.3h (1.94)	6.6i (2.67)	1.66ij
Turmeric	12.0c (3.5)	16.3b (4.09)	24.3b (4.97)	44.6b (6.86)	62.6b (7.94)	75.0b (8.68)	38.81bc
Kaolinite	0.0k (0.70)	1.3j (1.34)	6.0g (2.54)	10.0j (3.24)	12.3i (1.54)	18.6g (4.37)	8.04h
Bentonite	2.0j (1.58)	2.3i (1.67)	8.0f (2.91)	14.3i (3.84)	20.0e (4.52)	30.0f (5.52)	12.76g
Ash (cow dung)	12.6bc (3.61)	14.6cd (3.88)	22.6b (4.80)	30.0f (5.52)	53.3c (7.33)	68.3c (8.29)	33.57de
Sand	8.3e (2.96)	10.0f (3.24)	16.0cd (4.06)	29.3f (5.46)	40.3d (6.38)	55.6d (7.48)	25.58f
Saw dust	13.3b (3.71)	15.3c (3.97)	25.6ab (5.10)	44.0c (6.67)	63.3b (7.98)	78.3ab (8.87)	39.97ab
Malathion	0.0k (0.70)	0.0k (0.70)	0.0h (0.70)	3.0k (1.87)	5.33g (2.41)	9.0i (3.08)	2.88ij
Untreated control	15.0a (3.93)	19.3a (4.44)	28.6a (5.39)	52.6a (7.28)	68.3a (8.29)	85.0a (9.24)	43.13a
CD (1%)	0.14	0.10	0.32	0.16	0.32	0.45	2.80
SEm±	0.37	0.03	0.82	0.41	0.08	0.12	0.97

DAS – Days after storage   Means followed by the same letter do not differ significantly by DMRT (P=0.01)   Figures in parentheses are Öx + 0.5 transformed values mortality of *S. oryzae* in sorghum grains, treated with 1% sweet flag powder. Sunil Kumar (2003) and Yevoor (2003) reported that sweet flag powder at 1% caused cent per cent adult mortality of *S. oryzae* on maize grains at 90 days after storage (Sunil kumar, 2003).

Neem seed powder and custard seed powder both at 5% recorded 8.11 and 6.50 per cent grain damage and 2.2 and 2.1 per cent weight loss, respectively at 180 days after storage. These results are in conformity with Ali *et al* (1981) and Yevoor (2003).

Among different inert materials, kaolinite clay at 10% was found effective and recorded 92.3% adult mortality at 30 days after storage and it declined as the period of storage increased to 180 days after storage (75% mortality). The next best inert material was found to be bentonite clay which recorded mean grain damage and weight loss of 8.76, 0.22 per cent, respectively. These results are in conformity with Swamiappan (1978) who reported that the mortality is due to the removal of fat molecules from the cuticular waxy layer by absorption by clay particles resulting in water loss and subsequent death due to desiccation. However, the plant products were found superior over malathion.

References

Ali, S. I., Ravat, R. R. and O. P Singh. 1981. Effectiveness of various forms of custard apple seed powder as seed protectants against the pulse beetle, *Callosobruchus chinensis* L. *Bull. of Grain Tech.*, 19: 53-55.

Singh, K., Agarwal, N. S. and G. K Girish.. 1975. Studies on the damage of *Sitophilus oryzae* Linn. (Coleoptera:Curculionidae) and its effect on the germination of seeds of various high yielding varieties of wheat and maize *Indian J. Ent.*, 37 : 146-149.

Sivasrinivasu. 2001. Evaluation of indigenous products and boric acid against stored grain pests.*M.Sc. (Agri.) Thesis*, University of Agricultural Sciences, Dharwad.

Sunilkumar. 2003. Survey of indigenous technologies and evaluation of botanicals against major storage pests. *M. Sc. (Agri.) Thesis*, University of Agricultural Sciences, Dharwad.

Swamiappan, M., S. Jayaraj, K.C. Chandy and V. T Sundaramurthy. 1976. Effect of activated kaolinitic clay on some storage insects. *Zeitschrift fur Angewardte Entomologie*, 80:385-389.

Yevoor, 2003. Biology and management of rice weevil, *Sitophilus oryzae* (Linn.) in maize grains M.Sc. (Agri.) Thesis, University of Agricultural Sciences, Dharwad.

— xxx —

26

Indigenous Stored Grain Pest Management Practices in Kanyakumari District, Southern Peninsular India

KIRUBA, S.[1], S. JEEVA[2], S. ISRAEL STALIN[1] AND S. SAM MANOHAR DAS[1]

[1]Dept. of Post Graduate Studies and Research Centre in Zoology, Scott Christian College, Nagercoil – 629 003, TN, India

[2]Centre for Advanced Studies in Botany, School of Life Sciences, North-Eastern Hill University, Shillong – 793 022, India

SUMMARY

An investigation was undertaken to document the indigenous post harvest technologies adopted in stored products. Documentation of indigenous stored grain pest management practices showed that sun drying of seeds is an important post harvest pest management method followed in this tract. Products of plants like *Acorus calamus* (Rhizomes), *Capsicum frutescens* (fruits) and *Azadirachta indica* (leaves), *Pongamia pinnata* (leaves), *Psidium guajava* (leaves) and *Vitex negundo* (leaves) are being used as insect deterrents to prevent or deter insect infestation during storage. The stored produce in storage structure is topped up with fine soil or ash to manage certain pests.

Keywords: Ethnic Storage, Seed Purpose, Insect Deterrents.

1. INTRODUCTION

Proper stored produce pest management is essential to protect valuable grains from infestation by insects and other agencies (Kiruba *et al*, 2007). Food grains are being spoiled after harvest due to lack of sufficient storage and processing facilities (Singh and Satapathy, 2003; Jeeva *et al*, 2006). The annual losses in stored produce is estimated at 10 percent of all stored grain- about 13 million tonnes of grain loss due to insects alone (Wolpert, 1966). Damage by insects, fungi and sprouting causes loss of hundreds of millions of dollars worth economic damage to grain producers, merchandisers and processors each year (Harein and Meronuck, 1995).

Management of stored produce pests through fumigation (Page and Lubatti, 1963), chemical pesticides (Lemon, 1967) and plant-based deterrents (Schmutterer, 1990), saves only a portion of the produce from damage. Farm folk of the tropics and subtropics mainly follow traditional storage methods. These methods have a scientific basis, which is seldom to the users (Kiruba *et al*, 2006a, b).

People of Kanyakumari district, Tamil Nadu, India, follow time-tested customs and practices in various areas like food, medicines and agriculture. In this study the strategies followed by the people of Kanyakumari area for preserving stored produce have been exclusively studied.

2. MATERIALS AND METHODS

Kanyakumari the southern most district of Tamil Nadu, is described as a microcosm of the State, owing to its mosaic and diverse geographical and physical features such as lofty mountains and low plains, rivers and cascades, seacoast and thick inland forest, Red soils and fertile alluvium, a wide variety of flora, fauna, and protect the wild life. The Kanyakumari District is located in the world map, between $8^{o}03^{1}$ and $8^{o}35^{1}$ of the northern latitude and $77^{o}05^{1}$ to $77^{0}36^{1}$ of the eastern longitudes. The total geographical area of the district is 6,823 sq. km. The State of Kerala, Gulf of Mannar and the districts of Tirunelveli, and Thoothukudi surround the district.

The locality was visited and information was gathered about the prevalent stored produce pest management strategies through interviews. Identity of plants specimens was ascertained with the reference herbarium of the Botany Department, Scott Christian College, Nagercoil. The insect pests were identified with the reference specimens available in the Department of Zoology, Scott Christian College, Nagercoil. The grains and other products stored in the different structures were examined to ascertain their quality and pest infestation status.

3. RESULTS AND DISCUSSION

The traditional stored produce pest management strategies of the farmers of Kanyakumari District may be conveniently classified as biological methods and physicochemical methods.

3.1. Biological methods

Farmers use the following plants for achieving biological control of stored grain pests: *Cymbopogon citratus* Stapf. *Cymbopogon nardus* Rendl. *Pongamia glabra* Vent *Azadirachta indica* A.Juss. *Eucalyptus globulus* Labill., *Erythrina indica, Erythrina variegata* Linn., *Psidium guajava* L. *Vitex negundo.* L. and *Tephrosia purpurea* Pers.

The plants used by the farmers have already been reported to possess active components active against the insect pests. Srivastava *et al.* (1988) have reported that 0.2 percent (v/w) ginger grass oil on red gram prevented oviposition and F1 emergence of *chinensis* (L) for a long period after initial release of adults. Rehm and Espig (1991) established the high geraniol content of ginger grass oil. Paranagama *et al.*, (2003a) reported that damage to grain was lower in *C. citratus* and *C. nardus* treated *Oryza sativa*. These grasses showed deleterious effects on oviposition and F1 adult emergence of cow-pea bruchid, *C.maculatus* (F.) compared to the control during no-choice tests (Paranagama *et al.*, 2003b).

Storage structures are usually filled with dried leaves of *Pongamia glabra* and *A. indica* leaves applied to a filter paper exhibited repellency of 42 percent against *Tribolium castaneum* (Herbst) eight weeks after treatment, compared with 81.5 percent repellency, one week after treatment. Babu *et al.*, (1989) reported the inhibitory action of *P. glabra* seed oil- 0.75 percent (v/w) - on adult emergence of *C. chinensis* in pigeon pea. Dunkel *et al.* (1991) showed that the extract of *A. indica* seeds (0.2 percent w/w) applied to wheat caused 50 percent mortality in adult *Sitophilus oryzae* (L.) and 15 percent mortality in *Rhyzopertha dominica* (Fabricius) within three days and F1 emergence was reduced by 98 and 94 percent respectively. Khaire *et al.* (1992) reported *P. glabra* seed oil (5ml/kg) applied to wheat held in a store house for 12 months prevented natural infestations by the insects.

P. guajava and *V. negundo* leaves admixed with freshly harvested paddy, field infested with *Sitotroga cerealella*, significantly reduced the number of emerging F1 adults during four months of storage (Dakshinamurthy, 1988). The LD50 for *P. guajava* leaf powder admixed with rice assessed for Sitophilus oryzae L.and *Sitophilus granarius* at seven days was 2.25 percent and 2.28 percent (w/w) leaf powder, which prevented the production of F1 adults of both species (Sharaby, 1989). Morallo-Rejesus *et al*, (1990) showed that leaves of *V. negundo* caused 80 percent mortality in adult *C. chinensis* within 48h and prevented egg laying in mung bean. Prakash *et al.* (1990) showed that Z-heptatriacontanone isolated from the leaves of *Vitex negundo*, when admixed with rice at 400mg per Kg reduced oviposition in *Sitotroga cerealella, Rhyzopertha dominica* and *S. oryzae.*

3.2. Physicochemical methods

The following physicochemical methods were recorded during the present study - sun drying before storing the grains and application of fly ashes and dry sifted red soil.

The grains to be stored are sun dried by the farmers and this kills most of unwanted agents. Lale (1998) reported decreased oviposition and increased adult mortality of *Callosobruchus maculatus* in grains stored after exposure to sun. Lale and Vidal (2000) recorded 100 percent mortality of the eggs and first instar larvae of *C. maculatus* and first instar larvae of *Callosobruchus subinnotatus* (Pic) exposed to sunlight for 2 h in *Vigna subterranean* (L). Exposure of *C. maculatus* and *C. subinnotatus* adults to solar heat decreases the oviposition, retards egg development and reduces survival rate of immature stages. At a temperature of 50^0C, both species laid significantly fewer eggs with lower hatching potential, than those maintained at 40^0C. In both the species, no adult progeny emerged from seeds harbouring first instar larvae, when exposed to a temperature of 50^0C for 2,4 or 6h (Lale and Vidal, 2003). Exposure to the sun (in different coloured polyethylene bags) for 24h killed *chinensis* eggs and grubs found in infested green gram. Complete egg mortality was recorded in coloured bags and complete grub mortality, only in black coloured bags (Singh and Sharma, 2003).

The stored produce treated with dry sifted red soil, does not allow any infestation to the grains stored underneath. Normally the red soil layer is about 2 to 3 cm thick. Dry sifted red soil dust served as a protectant when applied to maize stored in bags (Le Pelly and Kockum,

1954). Such mineral dusts scratch the thin waterproofing layer of wax, which exists on the outside surface of the insect cuticle, allowing loss of water, which leads to death as a result of desiccation.

Hall (1970) reported the admixture of wood ash or sand with food grains in many areas, usually restricted to the storage of small quantities for seed purposes. Hakbigl (2002) reported the use of ash from burnt cow dung as an insecticide against *T. castaneum*, *T. granarius* (L.) and *Cryptolestes ferrugineus* (Stephens) larvae. Ashes kill insects by desiccation or by packing the inter granular spaces, restricting insect movement and emergence.

The use of' commonly available plant materials and other products help to reduce the huge financial inputs associated with chemical pesticides. They further prevent pollution of the environment, rendering the products pesticide - free, quite safe for human consumption.

Acknowledgement

The authors are grateful to the Head, Department of Zoology for extending the necessary facilities for successful completion of this study.

References

Babu, T.R., V.S. Reddy, and S.H. Hussaini, 1989. Effect of edible and non-edible oils on the development of the pulse beetle (*Callosobruchus chinensis* L.) and on viability and yield of mungbean (*Vigna radiata* L. Wilczek). *Tropical Science*, 29: 215-220.

Dakshinamurthy, A. 1988. Effect of certain plant products on storage pests of paddy. *Tropical Science*, 28: 119-122.

Dunkel, F.V., Sriharan, S., Niziyimana, E. and Serugendo, A. 1991. Evaluation of neem kernel extract (Margosan-O) against major stored insect pests of beans and sorghum in Rwanda. In: Proceedings of the Fifth International working conference on Stored -Product Protection, Fleurat-Lessard F & Ducom P, (eds.) Boreaux, France,September 1990, pp. 527-536.

Hakbijl, T. 2002. The traditional, historical and prehistoric use of ashes as an insecticide, with an experimental study on the insecticidal efficacy of washed ash. *Environmental Archaeology*, 7: 13-22.

Hall, D.W. 1970. Handling and storage of food grains in tropical and subtropical areas. FAO Agricultural Development paper No.90.

Harein, P. and Meronuck, R. 1995. Stored grain losses due to insects and molds and the importance of proper grain management. In: Stored Product

Management, (Krischik V, Cuperus G & Galliart D, eds.), E-912, CES Division of Agricultural Science Natural Research, OSU, USDA, FGIS, ES, APHIS, pp. 31.

Jeeva, S., Laloo, R.C. and Mishra, B.P. 2006. Traditional agricultural practices in Meghalaya, northeast India. *Ind. J. Tradi. Knowl.*, 5(1): 7–18.

Khaire, V.M., Kachare, B.V. and Mote, U.N. 1992. Efficacy of different vegetable oils as grain protectant against pulse beetle, *Callosobruchus chinensis* L. in increasing storability of pigeon pea. *J. Stored Prod. Res.*, 28: 153-156.

Kiruba, S., Mishra, B.P., Israel Stalin, S., Jeeva, S. and Sam Manohar Das, S. 2006a. Traditional pest management practices in Kanyakumari district, southern peninsular India. *Ind. J. Trad. Knowl.*, 5 (1): 71-74.

Kiruba, S., Sam Manohar Das, S. and Papadopoulou, S. 2006b. Prospects of traditional seed storage strategies against insect infestation adopted by two ethnic communities of Tamil Nadu, Southern Peninsular India. *Bull. Insect.*, 59(2): 129-134.

Kiruba, S., Jeeva, S., Das, S.S.M. and Kannan, D. 2007. Bamboo seeds as a means to sustenance of the indigenous community. *Ind. J. Trad. Knowl.*, 6 (1): 199-203.

Lale, N.E.S. 1998. Preliminary studies on the effect of solar heat on oviposition, development and adult mortality of the cowpea bruchid, *Callosobruchus maculatus* (F.) in the Nigerian Savanna. *J. Arid Env.*, 40 (2): 157-162.

Lale, N.E.S. and Vidal, S. 2000. Mortality of different development stages of *Callosobruchus maculatus* F. and *Callosobruchus subinnotatus* Pic. (Coleoptera: Bruchidae) in bambara groundnut, *Vigna subterranea* (L.) Verdcourt seeds exposed to simulated solar heat. *Zietschrift fiir Pflanzenkrankheiten und Pflanzenschutz*, 107 (5): 553-559.

Lale, N.E.S. and Vidal, S. 2003. Simulation studies on the effects of solar heat on egg laying, development and survival of *Callosobruchus maculatus* (F.) and *Callosobruchus subinnotatus* (Pic) in stored bambara groundnut, *Vigna subterranea* (L.) Verdcourt. *J. Stored Prod. Res.*, 39 (5): 447-458.

Le Pelly, R. and Kockum, S. 1954). Experiments in the use of insecticides for the protection of grain in storage. *Bull. Ent. Res.*, 45: 311.

Lemon, R.W. 1967. Laboratory evaluation of malathion bromophos and fenitrothion for use against beetle infesting stored products. *J. Stored Prod. Res.*, 2: 197-210.

Morallo-Rejesus, M.B., Maini, H.A., Hsawa, K. and Yamamota, I. 1990. Insecticidal actions of several plants to *Callosobruchus chinensis* L. In: Bruchids and legumes: Economics, Ecology and Co evolution, Fujii, K., Gatehouse, A.M.R., Johnson, C.D., Michel, R. and Yoshida, T. (eds.), Kluwer Academic Publishers, Dordrecht, The Netherlands, pp. 100.

Page, A.B.P. and Lubatti, O.F. 1963. Fumigation of insects. *Ann. Rev.Entomol*, 8: 239-264.

Paranagama, P., Abeysekera, T., Nugaliyadde, L. and Abeywickrama, K. 2003a. Effect of the essential oils of *Cymbopogon citratus*, *Cymbopogon nardus* and *Cinnomomum zeylanicum* on pest incidence and grain quality of rough rice (paddy) stored in an enclosed seed box. *J. Food, Agri. Env.*, 1(2): 139.

Paranagama, P., Adhikari, C., Abeywickrama, K. and Bandoura, P. 2003b. Deterrent effects of some Sri Lankan essential oils on oviposition and progeny production of the cowpea bruchid, *Callosobruchus maculatus* (F.) (Coleoptera: Bruchidae). *J. Food, Agri. Env.*, 1(2): 257.

Prakash, A., Mahapatra, P.K. and Mathur, K.C. 1990. Isolation, identification and evaluation of an oviposition inhibitor for storage insect pests from Begunia, *Vitex negundo* Linn. (Verbenaceae). *Bull. Grain Tech.*, 28 (1): 33-52.

Rehm, S. and Espig, G. 1991. The cultivated plants of the tropics and sub tropics. Verlag Josef Margraf. Weikersheim, Germany.

Schmutterer, H. 1990. Properties and potential of natural pesticides from the neem tree, *Azadirachta indica*. *Ann. Rev.Entomol.*, 35: 271-297.

Sharaby, A. 1989. Some Myrtaceae leaves as protectants of rice against the infestation of *Sitophilus oryzae* (L.) and *Sitophilus granarius* (L.) (Coleoptera), Polskie Pismo Entomologiczne. *Bulletin Entomologique de Pologne*, 59: 77-382.

Singh, R.K.P. and Satapathy, K.K. 2003. Zero Energy cool chamber: a low cost storage structure. *Journal of the North Eastern Council*, 23(3): 30.

Singh, S. and Sharma, G. 2003. Pulse beetle management through solar energy in stored green gram seeds. *Seed Res.*, 31(1): 89.

Srivastava, S., Gupta, K.C. and Agrawal, A. 1988. Effect of plant product on *Callosobruchus chinensis* L. infection on red gram. *Seed Res.*, 16(1): 98-101.

Wolpert, V. 1966. Rats in store, India. Far Eastern Economic Review, 53(10): 439.

— xxx —

27

Effect of Temperature and Relative Humidity on Efficacy of Diatomaceous Earth (DE) Against Pulse Beetle, *Callosobruchus chinensis* (L.)

POORNIMA MATTI[1] AND J.S.AWAKNAVAR[2]
[1] Research Scholar [2] Professor
Department of Agricultural Entomology, University of Agricultural Sciences
Dharwad -580005

SUMMARY

Efficacy of Diatomaceous Earth (DE) against the pulse beetle increased with decrease in relative humidity and increased temperature. The temperature and RH interaction revealed 100 per cent mortality even at highest RH (90%) at both higher temperatures (30 and 35°C) tested. There was slight decrease in the mortality at 20°C and 25°C with 70 per cent RH compared to lower RH combinations during storage period of two months. Germination of treated seeds ranged from 50 to 92 per cent witnessed in different combinations of temperature and RH tested.

Key words : Diatomaceous Earth, Environmental Hazard, Survival.

1. INTRODUCTION

Pulses constitute an integral part of Indian agriculture because they enrich human diet and soil fertility. Pulses contain 20-30 per cent protein of which lysine is of great importance and serves as the best means of solving malnutrition problems in the pure vegetarian Indian diet. In storage, a total of 25 species of insects have been recorded on pulses (Prabhakar, 1979), of them species of *Callosobruchus* belonging to family Bruchidae are very serious, *Callasobruchus chinensis* and *Callasobruchus maculatus* (F.) are predominant causing heavy damage to pigeonpea (Southgate, 1979). The loss in pulses during post harvest handling and storage was found at 8.5 per cent in India (Pingale et al., 1956) and zero per cent germination in greengram was recorded due to, *C. chinensis* infestation after six months of storage.

Indiscriminate use of insecticides has posed several problems like residual toxicity, resistance development and environmental hazards. There is a need to find the alternatives to the chemicals that can effectively prevent the storage losses, safer to the human beings and least detrimental to environment. This strategy could be achieved by exploiting formulated Diatomaceous Earth (DE). Diatomaceous earth is now very promising alternative to traditional pesticides in stored product protection (Korunic, 1998; Subramanyam and Roseli, 2000; Arthur, 2002). DEs are naturally occurring compounds, i.e., fossil of phytoplankton (diatoms) and their mode of action is purely physical: the DE particles adhere to the insect cuticle and absorb epicuticlar lipids, causing death by dessication (Korunic, 1998; Subramanyam and Roseli, 2000. Several DE formulations are now commercially available many of them have been evaluated with success against stored -grain insect pests.

In India no work on DE has been made. DE is generally known to lose its activity at higher relative humidity. In order to know the effect of prevailing environmental conditions, an experiment was conducted at the Department of Agricultural Entomology, UAS, Dharwad, to study the effect of temperature and humidities on the efficacy of the DE (Protect-It), against adults of *C. chinensis.*

2. MATERIALS AND METHODS

A formulated product of DE, "Protect-It" (M/s. Heldey Technologies Inc.USA) is a mixture of fresh water DE with 10 per cent silica aerogel to enhance insecticidal activity (Korunic and Fields, 1995). It is a beige-coloured DE containing approximately 83.70 per cent amorphous silicon dioxide, 5.6 per cent AI2O3, 2.3 per cent Fe2O3, 0.9 per cent CaO, 0.3 per cent MgO and 0.9per cent other oxides (e.g., TiO3, PJOJ) and 3 to 5 per cent moisture content. The median particle size is 5 jim. It contains significantly less than one per cent (usually less than 0.3%) crystalline silica content. This product was evaluated at a dose of 1 g per kg of seeds concentration under different levels of temperature and relative humidity. Constant temperature was maintained in the BOD incubator with a fluctuation of + 0.5 °C. The respective humidity levels were maintained using various saturated salt solutions (Winston and Bates, 1960) in the air- tight desiccators as follows.

The moisture content of the grains at the start of the experiment was determined. The desired dosage of the formulated DE was thoroughly mixed with 100g of uninfested red gram seeds in 100 ml beaker. Freshly emerged 1 to 2 day- old pulse beetles were drawn from the stock culture and released at the rate of 20 adults per beaker separately. The beetles were confined to the container by covering its mouth with muslin cloth fastened with rubber band. Again freshly emerged pulse beetles were released at every 15 days to see the efficacy.

3. RESULTS AND DISCUSSION

Results revealed 100 per cent mortality even at highest RH (90%) at both higher temperatures (30 and 35°C) tested, and at combinations of 20°C and 25°C with only lower RH (30 and 35 %). There was slight decrease in the mortality in both 20°C and 25°C with 70 per cent RH during storage period of two months compared to lower (Tables 1a,1b and 2). Mortality ranged around less than 10 per cent in untreated control. The present findings are in line with Rohita Prasantha *et al.* (2002) who reported that bruchids were exposed to the diatomaceous earth formulations. Fossil shield on treated mung beans and common beans at the rate of 1020 mg per kg and 1080 mg per kg, respectively. Diatomaceous earth treated bruchids were held at 20°C, 25°C, 30°C, 35°C temperature and 43, 52, 64, 75 and 84 per cent relative humidity, respectively. Mortality of bruchids generally increased as temperature increased but reverse was true with relative humidity in both the doses. Arthur (2002) reported that hard red winter wheat when treated with 300 ppm of Protect-It formulation of DE for one week, at combinations of 22°C, 27°C or 32°C, 40, 57 and 75 per cent relative humidity. No weevils survived when exposed at 40 or 57 per cent relative humidity and survival was related to both population density and temperature. More survival was noticed at 75 per cent relative humidity. Arthur (2001) revealed that *T. confusum* and *T. castaneum* exposed to the 0.5 mg per cm^2. Protect-It, met with increased mortality noticed at 27°C and 40 per cent relative humidity.

Table 1a : Details of treatments

Temperature (°C) Tested	Relative Humidity(%)	Salts
20	30	MgCl2 6H2O
	50	Mg(NO3)2 6H2O
	70	NaNO3
	90	MgSO4 7H2O
25	30	MgCl2 6H2O
	50	Mg(NO3)2 6H2O
	70	NaNO3
	90	BaCl2
30	30	MgCl2 6H2O
	50	Mg(NO3)2 6H2O
	70	NaNO3
	90	KN03
35	30	MgCl2 6H2O
	50	Mg(NO3)2 6H2O
	70	NaNO3
	90	KNO3

Highest seed germination (91 and 92 %) recorded at 30 DAT in 30°C and 35°C at 50 per cent RH and found to be significantly superior over other treatments. Whereas, 83 to 85per cent seed germination was recorded in 20°C and 25°C at 50% RH were the next best treatments. However, lowest seed germination was recorded in 35°C at 90% RH (Table 3). At 60 DAT, 30°C at 50% RH recorded highest germination of 90 per cent and found to be significantly superior over other treatments (Table 3). Whereas, 35°C at 50% RH (87%) was the next best treatment. However, lowest seed germination was recorded in 35°C at 90% RH. Highest seed germination was noticed due to less seed damage. The study reveals efficacy of diatomaceous earth in the protection of pulse seeds against pulse beetle.

Table 1b : Effect of levels temperature and relative humidity on efficacy of DE on mortality of *C. chinensis* (first release) in red gram % mortality at

Treatments	Dosage (%) (w/w)	1 DAT	3 DAT	5 DAT	7 DAT	Mean
T1-20°C at 30% RH	0.1	100.00a (89.96)	100.00a (89.96)	100.00a (89.96)	100.00a (89.96)	100.00a (89.96)
T2-20°C at 50% RH	0.1	100.00a (89.96)	100.00a (89.96)	100.00a (89.96)	100.00a (89.96)	100.00a (89.96)
T3-20°C at 70% RH	0.1	75.00e (59.99)	80,00e (63.44)	85.00d (67.63)	95.00c (71.31)	83.75f (66.20)
T4-20°C at 90% RH	0.1	65.00f (53.72)	75.00e (59.99)	SO.OOe (63.44)	90.00c (71.62)	77.50g (61.66)
T5-25°C at 30%RH	0.1	100.00a (89.96)	100.00a (89.96)	100.00a (89.96)	100.00a (89.96)	100.00a (89.96)
T6-25°C at 50% RH	0.1	100.00a (89.96)	100.00a (89.96)	100.00a (89.96)	100.00a (89.96)	100.00a (89.96)
T7-25°C at 70% RH	0.1	80.00d (63.41)	90.00c (71.54)	95.00b (77.05)	95.00b (77.05)	91.25d (72.77)
T8-25°C at 90% RH	0.1	75.00e (59.98)	85.00d (67.19)	90.00c (71.56)	97.50b (81.06)	86.87e (68.73)
T9-30°C at 30%RH	0.1	100.00a (89.96)	100.00a (89.96)	100.00a (89.96)	100.00a (89.96)	100.00a (89.96)
T10-30°C at 50% RH	0.1	100.00a (89.96)	100.00a (89.96)	100.00a (89.96)	100.00a (89.96)	100.00a (89.96)
T11- 30°C at 70% RH	0.1	90.00b (71.54)	95.00b (77.05)	100.00a (89.96)	100.00a (89.96)	96.25b (78.90)
T12-30°C at 90%RH	0.1	85.00c (67.19)	90.00c (71.54)	95.00c (71.54)	100.00a (89.96)	92.50c (74.11)
T13-35°C at 30% RH	0.1	100.00a (89.96)	100.00a (89.96)	100.00a (89.96)	100.00a (89.96)	100.00a (89.96)
T14-35°C at 50% RH	0.1	100.00a (89.96)	100.00a (89.96)	100.00a (89.96)	100.00a (89.96)	100.00a (89.96)
T15-35°C at 70%RH	0.1	100.00a (89.96)	100.00a (89.96)	100.00a (89.96)	100.00a (89.96)	100.00a (89.96)
T16-35°C at 90% RH	0.1	100.00a (89.96)	100.00a (89.96)	100.00a (89.96)	100.00a (89.96)	100.00a (89.96)

	1 DAT SEm+	1 DAT CD	3 DAT SEm+	3 DAT CD	5 DAT SEm+	5 DAT CD	7 DAT SEm+	7 DAT CD	Mean SEm+	Mean CD
RH	0.16	0.49	0.32	0.97	0.27	0.82	0.33	1.00	0.21	0.64
Temp.	0.14	0.44	0.29	0.88	0.25	0.75	0.30	0.91	0.19	0.59
RH x Temp.	0.40	1.22	0.47	1.42	0.57	1.61	0.86	2.60	0.42	1.28

DAT - Days after treatment
Figures in parentheses are arc sine transformed values
Means in the columns followed by the same alphabet do not differ significantly by DMRT (P=0,01%)

Table 2 : Effect of levels of temperature and relative humidity on efficacy of DE on mortality of *C. chinensis* (fourth release) in red gram % mortality at

Treatments	Dosage (%) (w/w)	1 DAT	3 DAT	5 DAT	7 DAT	Mean
T1-20°C at 30% RH	0.1	100.00a (89.96)	100.00a (89.96)	100.00a (89.96)	100.00a (89.96)	100.00a (89.96)
T2-20°C at 50% RH	0.1	100.00a (89.96)	100.00a (89.96)	100.00a (89.96)	100.00a (89.96)	100.00a (89.96)
T3-20°C at 70% RH	0.1	SO.OOe (63.44)	100.00a (89.96)	100.00a (89.96)	100.00a (89,96)	96.25d (78.80)
T4-20°C at 90% RH	0.1	75.00f (59.99)	90.00c (71.56)	92.50c (74.08)	100.00a (89.96)	90.65e (72.19)
T5-25°C at 30%RH	0.1	100.00a (89.96)	100.00a (89.96)	100.00a (89.96)	100.00a (89.96)	100.00a (89.96)
T6-25°C at 50% RH	0.1	100.00a (89.96)	100.00a (89.96)	100.00a (89.96)	100.00a (89.96)	100.00a (89.96)
T7-25°C at 70% RH	0.1	90.00c (71.54)	100.00a (89.96)	100.00a (89.96)	100.00a (89.96)	98.75b (84.22)
T8-25°C at 90% RH	0.1	85.00c (67.23)	92.50b (74.39)	95.00b (77.31)	100.00a (89.96)	94.37c (76.24)
T9-30°C at 30% RH	0.1	100.00a (89.96)	100.00a (89.96)	100.00a (89.96)	100.00a (89.96)	100.00a (89.96)
T10-30°C at 50%RH	0.1	100.00a (89.96)	100.00a (89.96)	100.00a (89.96)	100.00a (89.96)	100.00a (89.96)
T11-30°C at 70% RH	0.1	95.00b (77.05)	100.00a (89.96)	100.00a (89.96)	100.00a (89.96)	100.00a (89.96)
T12-30°C at 90% RH	0.1	90.00c (71.64)	100.00a (89.96)	100.00a (89.96)	100.00a (89.96)	100.00a (89.96)
T13-35°C at 30% RH	0.1	100.00a (89.96)	100.00a (89.96)	100.00a (89.96)	100.00a (89.96)	100.00a (89.96)
T14-35°C at 50% RH	0.1	100.00a (89.96)	100.00a (89.96)	100.00a (89.96)	100.00a (89.96)	100.00a (89.96)
T15-35°C at 70% RH	0.1	100.00a (89.96)	100.00a (89.96)	100.00a (89.96)	100.00a (89.96)	100.00a (89.96)
T16-35°C at 90% RH	0.1	100.00a (89.96)	100.00a (89.96)	100.00a (89.96)	100,00a (89.96)	100.00a (89.96)

		1 DAT SEm+	1 DAT CD	3 DAT SEm+	3 DAT CD	5 DAT SEm+	5 DAT CD	7 DAT SEm+	7 DAT CD	Mean SEm+	Mean CD
RH		0.32	0.97	0.35	1.06	0.27	0.82	-	-	0.31	0.93
Temp.		0.29	0.88	0.03	0.09	0.26	0.78	-		0.30	0.90
RH x Temp.		0.40	1.22	0.66	2.00	0.58	1.75	-	-	0.72	2.17

DAT - Days after

Figures in parentheses arc sine transformed values

Means in the columns followed by the same alphabet do not differ significantly by DMRT (P=0,01%)

Table 3. Effect of DE under various regimes of temperature and humidity on germination of redgram seeds

Treatments	DE Dosage (%) (w/w)	per cent seed germination at 30 DAT	60 DAT	Mean
T1-20°C at 30% RH	0.1	56.001 (48.43)	53.00i (45.70)	54.50h (47.56)
T2-20°C at 50% RH	0.1	83.00cd (65.63)	81.00cd (64.13)	82.00c (64.88)
T3 - 20°C at 70% RH	0.1	75.00f (60.64)	75.00ef (59.98)	75.50e (60.31)
T4 - 20°C at 90% RH	0.1	50.00J (44.98)	49.00J (44.41)	49.50i (44.69)
T5-25°C at 30% RH	0.1	59.00h (50.17)	57.00h (49.00)	58.00g (49.58)
T5-25°C at 50% RH	0.1	85.00c (67.19)	83.00c (65.62)	84.00b (65.41)
T6-25°C at 70% RH	0.1	77.00e (61.32)	75.00f (59.98)	76.00e (60.65)
T7-25°C at 90% RH	0.1	45.00jk (42.11)	43.00k (40.96)	44.00J (41.44)
T8-30°C at 30% RH	0.1	63.00g (52.52)	61.00g (51.33)	61.50J (51.63)
T9-30°C at 50% RH	0.1	91.00b (72.54)	90.00a (71.56)	90.50a (72.02)
T10-30°C at 70% RH	0.1	79.00d (62.71)	76.00e (61.32)	77.50de (61.66)
T11-30°C at 90% RH	0.1	41.00k (39.79)	40.00k (39.22)	40.500 (39.51)
T13-35°C at 30% RH	0.1	64.00g (53.11)	61.00g (51.34)	62.50f (52.22)
T14-35°C at 50% RH	0.1	92.00a (73.66)	87.00b (68.85)	89.50a (71.07)
T15-35°C at 70%RH	0.1	80.00d (63.41)	77.00e (61.32)	78.50d (62.35)
T16-35°C at 90% RH	0.1	41.00k (39.79)	40.00k (39.22)	40.50h (39.51)

	30 DAT SEm+	30 DAT CD	60 DAT SEm+	60 DAT CD	Mean SEm+	Mean CD
RH	0.31	0.93	0.23	0.71	0.23	0.70
Temp.	0.28	0.86	0.21	0.64	0.21	0.64
RH x Temp.	0.83	2.50	0.47	1.41	0.49	1.48

DAT - Days after treatment
Figures in parentheses are arc sine transformed values
Means in the columns followed by the same alphabet do not differ significantly by DMRT (P=0.01%)

References

Arthur, F. H., 2001. Impact of food source on survival of red flour beetles and confused flour beetles (Coleoptera:Tenebrionidae) exposed to diatomaceous earth. *J. Econ. Entomol.,* 93(4) : 1347-1356.

Arthur, F, H., 2002. Survival of Sitophilus oryzae (L.) on wheat treated with iatomaceous earth : impact of biological and environmental parameters on product efficacy. *J. Stored Prod. Res.,* 38 : 305-313.

Korunic, Z. 1998. Diatomaccous earth, a group of natural insecticides. *J. Stored Prod. Res.,* 34: 87-97.

Korunic, Z. and P. G Fields.1995, Diatomaceous earth insecticidal composition. Canadian and USA Patent.

Pingale, S. V., S. B.Kandkal, and M. Swaminathan,., 1956, Effect of insect infestation on stored bengalgram (*Cicer arietinum* L.) and greengram (*Phaseolus radiatus* L.). *Bull. CFTRI,* 5:211-213.

Prabhakar, G. S., 1979. Studies on the bruchid fauna infesting pulse crops of Karnataka with special emphasis on bioefficacy of *Callosobruchus chinensis* (L.) (Coleoptera : Bruchidae). M.Sc. (Agri.) Thesis, University of Agricultural Sciences, Bangalore, p. 147.

Rohitha Prasantha, B. D., C. H Reichrauth,. and C Buttner, 2002. Effect of temperature and relative humidity on diatomaceous earth treated *Callosobruchus tnaculatus* and *Acanthoscelides obtectus* (Coleoptera : Bruchidae). *Proceedings of the 8lh International Working Conference on Stored Product Protection,* 2 : 987-993.

Southgate, B. J., 1979, Biology of burchidae. *Ann. Rev. Entomol.,* 24 : 449-473.

Subramanyam, Bh., and R. Roseli. 2000. Inert dusts, pp321-380. In: Bh. Subramanyam and D.W.Hagstrum(Eds), Alternatives to pesticides in stored-product IPM. Kluwer, Dordreecht, The Netherlands.

Winston, P.W. and D.H. Bates, I960. Saturated salt solutions for biological research. *Ecology,* 41(1): 232-236.

— xxx —

28

Effect of Pre-harvest Sanitation Spray on Seed Yield, Quality and Post-harvest Seed Storability in Greengram *(Vigna radiata* (L.) Wilczek)

PATIL, R.H., M.N. MERWADE AND J. DIVYASHRI
AICRP on Soybean, Main Agricultural Research Station, University of Agricultural Sciences, Dharwad - 580 005

SUMMARY

Results showed that irrespective of seed treatments and storage containers, seed quantitative parameters of seeds like seed infection and seed weight showed consistent increasing trend while seed quality parameters like 100 seed weight, germination, germination rate index and seedling vigour exhibited significantly decreasing trend with the advancement of storage period upto 10 months.Irrespective of containers greengram seeds treated with castor oil, neem oil, malathion and sweet flag rhizome powder retained higher seed quality parameters with less quantitative losses during the 10 months of storage period. Similar results were also seen in the seeds stored in polythene bag over cloth bag during the entire storage period. In general, seeds treated with castor oil and stored in polythene bag retained higher seed quality by showing the minimum seed quantitative losses throughout the period.

Key words: Sanitation Spray, Seed Storability, Seed Germination.

1. INTRODUCTION

Among the several insect pests, pulse beetle or bruchids *(Callosobruchus chinensis* L.*)* is a serious storage pest causing major loss in greengram seeds as it is known to lay its eggs on seeds in the field itself even before harvest and becomes a menace during storage period. There is a need to preserve high quality of greengram seeds by controlling storage pests at both pre and post-harvest levels by using suitable insecticides and botanicals. Botanicals such as castor oil, neem oil, sweet flag rhizome powder are found more effective and safe in maintaining high quality of seeds in storage for longer period due to their oviposition inhibitory and repellent properties. In this backdrop, the present study was conducted to know the impact of pre and post harvest insecticidal applications on seed quality and storability of greengram seeds.

2. MATERIAL AND METHODS

A field experiment to test five insecticides (I) *viz.,* control (Io), malathion @ 2 ml/1 (Ii), methyl parathion @ 1 ml/1 (Ij), nimbicidine @ 5 ml/1 (I3) and Indian Privet leaf extract @ 5 ml/lit (I4) as main factor and three stages of insecticidal sprays ***viz.,*** 20 days before harvest (DBH) (SO, 10 DBH (S2) and 20+10-DBH (S3) as sub factors (Table 1) was conducted in randomized block design with two factorial concept in three replications at the Main Agricultural Research Station, Dharwad Ten plants chosen randomly were observed from each treatment from net plot. And a laboratory experiment was also conducted under ambient conditions for 10 months storage period with eleven seed treatments (Tables 2 and 3) as a main factor and two types of storage containers *viz.,* cloth bag (Ci) and 700 gauge polythene bag (C2) as sub factors. This experiment was laid out in completely randomized design with two factorial concept with three replications. Standard germination test was carried out adopting the between paper method.

3. RESULTS AND DISCUSSION

Significant variations on infected pods per plant were observed due to insecticides over stages of insecticidal sprays. Significantly less pod infestation (5.54%) was noticed in malathion followed by nimbicidine (6.25%) and are differed significantly with other treatments (Table-1a). Irrespective of the insecticides, stages of insecticidal sprays showed significant variations on infected pods per plant, spraying

Table 1a. Efficacy of insecticides and stages of their application on the infestation of *C. chinensis* in greengram pods

Treatments	Infested pods (%)				Seed yield (kg/ha)				Infested seeds per plant (%)			
	S_1 (20 DBH)	S_2 (10DBH)	S_3 (20+10 DBH)	Mean	S_1 (20 DBH)	S_2 (10 DBH)	S_3 (20+10 DBH)	Mean	S_1 (20 DBH)	S_2 (10 DBH)	S_3 (20+10 DBH)	Mean
I_0- Control (No insecticide)	13.80 (21.90)	14.10 (22.04)	14.50 (22.14)	14.13 (22.02)	735.0	715.0	745.0	731.7	7.55 (16.18)	7.87 (16.28)	8.10 (16.48)	7.84 (16.28)
I_1- Malathion @ 2 ml/litre	5.84 (13.97)	6.71 (15.00)	4.09 (11.66)	5.54 (13.54)	968.5	910.0	1004.5	961.0	2.67 (9.39)	3.53 (10.80)	1.79 (7.64)	2.66 (9.28)
I_2- Methyl parathion @ 1 ml/litre	7.00 (15.34)	8.0S (16.75)	6.26 (14.48)	7.11 (15.52)	888.0	857.0	900.0	881.7	4.75 (12.56)	5.51 (13.56)	3.46 (10.69)	4.57 (12.27)
I_3- Nimbicidine @ 5 ml/litre	6.22 (14.43)	7.34 (15.70)	5.19 (13.16)	6.25 (14.43)	949.9	900.0	985.0	945.0	3.59 (10.89)	4.00 (11.53)	2.82 (9.65)	3.47 (10.69)
I_4- Indian Privet leaf extract @ 5 ml/litre	8.03 (16.45)	9.24 (16.68)	7.15 (15.50)	8.14 (16.21)	848.0	817.0	869.0	844.7	5.37 (13.38)	6.40 (14.64)	4.25 (11.84)	5.34 (13.29)
Mean	8.24 (16.39)	9.09 (17.30)	7.36 (15.35)	8.23 (16.35)	877.8	840.0	900.7	872.3	4.85 (12.38)	5.46 (13.41)	4.03 (11.29)	4.78 (12.36)
For comparison of treatments means of		S.Em±	CD at 5%			S.Em±	CD at 5%			S.Em±	CD at 5%	
Insecticides (I)		0.19	0.55			11.15	32.32			0.15	0.45	
Sprays (S)		0.15	0.42			8.64	25.04			0.12	0.35	
IxS		0.33	NS			19.31	NS			0.27	0.78	

S_1- Foliar spray at 20 days before harvest
S_2- Foliar spray at 10 days before harvest
S_3- Foliar spray at 20 and 10 days before harvest

insecticides at 20 and 10 DBH (S3) recorded significantly less pod infestation (736%) compared to other stages. The interaction effects between the insecticides and stage of insecticidal sprays was found to be non-significant (Table 1a).

Significant variations in seed yield per hectare due to insecticides over stages of insecticidal sprays were recorded. Maximum seed yield per hectare was obtained (961.0 kg) in the malathion (Ii) which was on par with nimbicidine (I3) (945.0 kg). And it was differed significantly due to stages of insecticidal sprays over the insecticides. Spraying of insecticides at 20 and 10 DBH (S3) recorded significantly high seed yield per hectare (900.2 kg) compared to other stages. Whereas, it was significantly lower (877.9 kg) in the insecticidal spraying at 10DBH (S2) which was on par with insecticidal spraying at 20 DBH (Si) (840.0 kg) whereas interaction effect between treatments was non-significant.

Infested seeds per plant exhibited marked variations due to insecticides over stages of insecticidal sprays. Significantly less seed infestation (2.66%) was noticed in malathion (Ii) compared to the other treatments. The next lower seed infestation was obtained with nimbicidine (3.47%) (Table 1a). The stages of insecticide sprays showed significant differences on infested seed per plant irrespective of the insecticides,. Insecticidal spraying at 20 and 10 DBH (S3) recorded significantly less seed infestation (4.03%) and more (5.46%) at spraying of insecticides at 10DBH (Si), respectively compared to other stages.

The interaction effects between insecticides and stages of sprays was found significantly different for the infested seeds. Significantly the lowest (1.79%) and highest (8.10%) seed infestation was noticed in the treatment combinations of malathion @ 2 ml/lit sprayed at 20+10 DBH (I1S3) and untreated combination, respectively, compared to other treatment combinations. Irrespective of stages of insecticidal sprays, significant differences on the seed germination were seen due to different insecticides. Significantly the highest seed germination (89.70%) was recorded with malathion followed by nimbicidine (87.37%) and methyl parathion (85.55%) (Table 1b). Marked variations on seed germination were observed due to stages of insecticidal sprays over the insecticides. Spraying of insecticides at 20 and 10DBH recorded significantly higher seed germination (87.13%) compared to other stages. Whereas, it was significantly low (83.88%) in insecticidal spray given at 10 DBH which was on par with individual spraying at 20 DBH (84.94%). The interaction effects of insecticides and stages of spray were found non significant.

Table 1b: Efficacy of insecticides and stages of their application on germination of greengram seeds

Treatments	Germination (%)			
	S_1 (20 DBH)	S_2 (20 DBH)	S_3 (20 DBH)	Mean
I_0- Control (No insecticide)	78.64 (63.38)	79.18 (63.40)	79.94 (63.42)	79.24 (63.40)
I_1- Malathion @ 2 ml/litre	71.05 (86.13)	84.15 (69.30)	91.88 (71.28)	89.70 (71.37)
I_2- Methyl parathion @ 1 ml/litre	67.83 (85.76)	84.13 (66.52)	86.75 (68.67)	85.55 (67.67)
I_3- Nimbicidine @ 5 ml/litre	88.97 (70.60)	88.10 (69.86)	92.04 (73.63)	87.37 (70.52)
I_4- Indian Privet leaf extract @ 5 ml/litre	83.91 (66.35)	83.10 (65.72)	85.04 (67.25)	84.02 (66.44)
Mean	84.94 (67.85)	83.88 (66.96)	87.13 (68.84)	85.32 (67.88)
For comparison of treatments means of		S.Em±	CD at 5%	
Insecticides (I)		0.66	1.92	
Sprays (S)		0.51	1.49	
I x S		1.15	NS	

These results confirmed the usefulness of spraying of botanicals and insecticides at different stages of crop growth period by showing marked decrease in percentage of infected pods and infected seeds/ plant but increase in seed yield attributing components in greengram. These results also pointed out the beneficial effect of botanicals in obtaining the seeds of superior quality parameters and it might be related to increase in seed yield attributes like number of pods/plant, seed yield/plant with less number of infected pods and seeds/plant. These findings are in agreement with those of Patrick *et al.* (1998) in pea, Merwade (2000) in chickpea and Biradar (2001) in greengram.

3.1. Seed treatment and storage containers

3.1.1. Seed infestation

Seed infestation differed significantly among the seed treatments in all the ten months of storage period irrespective of storage containers used. At second month of storage period, castor oil (T2), neem oil (Ti), sweet flag rhizome powder (T3), custard apple powder (T4), malathion (T8) and methyl parathion (T9) showed nil infestation of seeds. Significantly lower seed infestation was recorded in red soil clay (T7) (0.3%) which was on par with chlorax (T10) (0.39%). Whereas, control

(T10) treatment recorded significantly higher (4.89%) seed infestation (Table 2). during fourth month of storage period seed infestation was zero in castor oil (T2), neem oil (Ti) and malathion (Tg) which was followed by sweet flag rhizome powder (T3) (0.28) and it remained on par with methyl parathion (T9) (0.42%). Whereas, it was significantly higher (17.46%) in control (T--1o). At sixth and eighth month period, castor oil (T2) recorded significantly lower (0.12% and 2.91%) which was on par with neem oil (0.25% and 3.54%) and it was significantly higher in control (To) (25.50 and 37.87) respectively. At the end of ten months storage period, seed infestation was markedly lowest in castor oil (8.30%) which was on par with neem oil (9.13%) but it differed significantly with rest of the seed treatments. Significantly highest infestation(59.53%) was noticed in the control (Table 2a).

The seed infestation was nil at initial stage in both the cloth bag (Ci) and 700 gauge polythene bag (C2). At the end of two, four, six and eight months of storage period, the seed infestation was significantly less in the polythene bag (C2) (0.38, 1.57, 3.52, 7.82%) and more in cloth bag (d) (0.97, 4.07, 7.32, 16.74%) respectively. At ten months storage period, the seed infestation was significantly less in polythene bag (C2) (18.23%) than in cloth bag (Cj) (31.30%). The interaction effect between seed treatment and containers (TxC) revealed insignificant variation on per cent seed infestation throughout the ten months storage period.

3.1.2. Seed germination

Irrespective of storage containers, seed germination showed significant variations throughout the storage period except in initial month (Table 2b). At initial month non significant variance on seed germination showed numerically higher and lower per cent germination in castor oil (T2) 92.05 and control (To) 92.00 respectively. At second, fourth, sixth and eight months storage period, significantly higher seed germination was seen in castor oil (T2) (91.34, 90.74, 87.46, 82.31) and it was on par with neem oil (T,) (91.00, 90.65, 86.67, 81.80%) and malathion (Tg) (90.34, 89.92, 86.59, 81.55%) where as it was significantly lower in (To) (87.90, 85.04, 81.64, 76.62%) respectively (Table 3).

The marked variations on germination were noticed among the containers in all the months of storage period except in initial and two month period over seed treatments. At initial and two months storage

Table 2a : Effect of seed treatments and storage containers on infestation of *C. chinensis* in greengram seeds in storage

Treatments		Seed infestation (%) at Periods of storage (months)				
	0	2	4	6	8	10
To - Control (No seed treatment)	0	4.89 (12.59)	17.46 (24.39)	25 50 (29 90)	37 87 (37.85)	59.53 (50.97)
T1 - Neem oil @ 5 ml per kg of seed	0	0.00 (0.25)	0.00 (0.25)	0.25 (2.05)	3.54 (10 51)	9.13 (17.24)
T2 - Castor oil @ 5 ml per kg ofseed	0	0.00 (0.25)	0.00 (0.25)	0.12 (1.49)	2.91 (9.40)	8.30 (16.40)
T3 - Sweet flag powder @ 10 g per kg of seed	0	0.00 (0.25)	0.28 (2.28)	2.39 (8.58)	8.39 (16 57)	22.08 (27.79)
T4 - Custard apple seed powder @ 10g per kg of seed	0	0.00 (0.25)	0.90 (5.01)	5.31 (13 34)	13.87 (21 39)	26.22 (30.52)
T5 - Soapnut seed powder @ 10 gper kg of seed	0	0.76 (4.57)	2.67 (8.85)	6.74 (14.71)	15.25 (22.45)	28.06 (31.74)
T6 - Wood ash powder @ 5 g per kg of seed	0	1.13 (5.59)	3.75 (10.56)	3.39 (10.27)	11.37 (19.33)	29.64 (32.75)
T7- Red soil clay @ 5 g per kg of seed	0	0.30 (2.30)	4.16 (11.33)	7.96 (16.14)	16.03 (23.02)	23.45 (28.68)
Tg - Malathion @ 5 g per kg of seed	0	0.00 (0.25)	0.00 (0.25)	1.46 (6.59)	5.60 (13.30)	16.11 (23.30)
T9 - Methyl parathion @ 5 g per kg of seed	0	0.00 (0.25)	0.42 (2.76)	2.94 (9.52)	9.52 (17.63)	23.69 (28.88)
T10 - Chlorax @ 5 g per kg of seed	0	0.39 (2.65)	1.43 (6.44)	3.27 (10.15)	10.73 (18.73)	26.17 (30.51)
S.Em±	NA	0.25	0.35	0.53	0.41	1.12
CD (5%)	NA	0.69	0.97	1.49	1.14	3.12
Containers (C)						
C_1-Cloth bag	0	(0.97) (3.59)	4.07 (8.65)	7 32 (63.65)	16.74 (23.17)	31.30 (35.58)
C_2 - Polythene bag (700 gauge)	0	0.38 (1.72)	1.57 (4.51)	3 52 (8 67)	7 82 (15.04)	18.23 (24.38)
Mean		0.6 (2.65)	2.82 (6.58)	5.42 (11.16)	12.28 (19.11)	24.76 (28.98)
S.Em±	NA	0.10	0.14	0.22	0.17	0.48
CD (5%)	NA	0.29	0.41	0.63	0.48	1.33

period, numerically higher germination was recorded in polythene bag (C2) (92.02, 89.97%) but it was less in cloth bag (Ci) (92.01 and 89.46%) respectively. At four, six and eight month storage period, polythene bag (C2) recorded significantly more seed germination of 89.64, 86.77 and 81.73 per cent but it was less in cloth bag (C|) (87.80, 84.04 and 78.74%) respectively. In ten months storage period seed germination was significantly higher (76.76%) in polythene bag (C2) than in cloth bag (d) (72.94%) (Table 3).

Table 2b : Effect of seed treatments and storage containers on infestation of *C. chinensis* in greengram seeds in storage

Interaction	Seed Germination (%) at Periods of storage (months)					
	0	2	4	6	8	10
T_0C_1	0	6.15 (14.30)	22.97 (28.58)	32.25 (34.21)	45.75 (42.50)	67.83 (56.24)
T_0C_2	0	3.63 (10.88)	11.95 (20.19)	18.75 (25.59)	30.00 (33.20)	51.24 (45.70)
T_1C_1	0	0.00 (0.25)	0.00 (0.25)	0.50 (3.86)	5.24 (13.22)	12.75 (20.90)
T_1C_2	0	0.00 (0.25)	0.00 (0.25)	0.00 (0.25)	1.85 (7.79)	5.52 (13.57)
T_2C_1	0	0.00 (0.25)	0.00 (0.25)	0.25 (2.74)	4.30 (12.23)	11.60 (19.90)
T_2C_2	0	0.00 (0.25)	0.00 (0.25)	0.00 (0.25)	1.32 (6.58)	5.00 (12.91)
T_3C_1	0	0.00 (0.25)	0.57 (4.31)	3.18 (10.21)	11.31 (19.63)	28.34 (32.15)
T_3C_2	0	0.00 (0.25)	0.00 (0.25)	1.60 (6.94)	5.47 (13.51)	15.82 (23.42)
T_4C_1	0	0.00 (0.25)	1.52 (7.00)	7.95 (16.37)	19.75 (26.33)	33.90 (35.55)
T_4C_2	0	0.00 (0.25)	0.28 (3.02)	3.27 (10.34)	8.00 (16.41)	18.57 (25.50)
T_5C_1	0	1.32 (6.56)	4.34 (11.99)	9.48 (17.90)	22.00 (27.96)	36.12 (36.93)
T_5C_2	0	0.21 (2.58)	1.00 (5.70)	4.00 (11.52)	8.50 (16.94)	20.00 (26.56)
T_6C_1	0	1.89 (7.76)	5.97 (14.09)	5.00 (12.91)	15.77 (23.38)	37.57 (37.75)
T_6C_2	0	0.37 (3.42)	1.54 (7.04)	1.79 (7.64)	6.98 (15.29)	21.72 (27.76)
T_7C_1	0	0.61 (4.35)	6.32 (14.54)	0.72 (19.09)	23.32 (28.87)	30.10 (33.18)
T_7C_2	0	0.00 (0.25)	2.00 (8.12)	5.21 (13.18)	8.74 (17.18)	16.80 (24.18)
T_8C_1	0	0.00 (0.25)	0.00 (0.25)	2.27 (8.58)	8.21 (16.63)	21.91 (27.89)
T_8C_2	0	0.00 (0.25)	0.00 (0.25)	0.65 (4.60)	3.00 (9.96)	10.32 (18.72)
T_9C_1	0	0.00 (0.25)	0.85 (15.28)	4.37 (12.05)	13.07 (21.18)	30.43 (33.4 7)
T_9C_2	0	0.00 (0.25)	0.00 (0.25)	1.51 (6.98)	5.98 (14.08)	16.95 (24.30)
$T_{10}C_1$	0	0.78 (5.06)	2.30 (8.62)	4.54 (12.18)	15.25 (22.97)	33.75 (35.48)
$T_{10}C_2$	0	0.00 (0.25)	0.57 (14.31)	2.00 (8.12)	6.27 (14.49)	18.60 (25.54)
Mean		0.68 (2.65)	2.82 (16.58)	5.42 (11.16)	12.28 (19.11)	24.76 (28.98)
S.Em±	NA	0.35	0.49	0.76	0.58	1.59
CD 5%	NA	NS	NS	NS	NS	NS

Table 3 : Effect of seed treatments and storage containers on infestation of *C. chinensis* in greengram seeds in storage

Treatments		Seed Germination (%) at Periods of storage (months)				
Seed treatments (T)	0	2	4	6	8	10
T_0 - Control (No seed treatment)	92.00 (73.55)	87.90 (69.34)	85.04 (67.26)	81.64 (64.65)	76.62 (61.09)	70.83 (57.32)
T_1- Neem oil @ 5 ml per kg of seed	92.04 (73.61)	91.00 (72.54)	90.65 (72.21)	88.57 (68.54)	81.80 (64.76)	77.12 (61.43)
T_2 - Castor oil @ 5 ml per kg of seed	92.05 (73.60)	91.34 (72,88)	90.74 (72.30)	87.46 (69.29)	82.31 (65.14)	77.40 (61.62)
T_3 - Sweet flag powder @ 10 g per kg of seed	92.04 (73.61)	90.37 (71.92)	89.22 (70.86)	86.16 (68.18)	81.10 (64.25)	74.10 (59.46)
T_4- Custard apple seed powder @ 10 g per kg of seed	92.03 (73.59)	90.10 (71.67)	89.07 (70.65)	85.47 (67.61)	80.50 (63.81)	74.83 (59.90)
T_5 - Soapnut seed powder @10 g per kg of seed	92.00 (73.55)	88.21 (69.92)	88.20 (69.92)	84.95 (67.19)	79.30 (62.94)	73.95 (59.33)
T_6- Wood ash powder @ 5 g per kg of seed	92.01 (73.59)	89.17 (70.78)	88.58 (70.26)	84.70 (66.98)	79.54 (63.11)	73.76 (59.19)
T_7- Red soil clay @ 5 g per kg of seed	92.00 (73.56)	90.04 (71.60)	88.31 (70.04)	85.93 (67.99)	80.58 (63.86)	75.0a (60.07)
T_8 - Malathion @ 5 g per kg of seed	92.03 (73.59)	90.34 (71.79)	89.92 (71.50)	86.59 (68.54)	81.55 (64.59)	76.62 (61.09)
T_9- Methyl parathion @ 5 g per kg ofseed	92.01 (73.57)	89.36 (89.36)	88.28 (70.04)	85.24 (67.42)	79.97 (63.44)	75.22 (60.29)
T_{10} - Chlorax @ 5 g per kg of seed	92.00 (73.53)	89.07 (89.07)	87.95 (69.70)	84.62 (66.91)	79.31 (62.95)	74.47 (59.66)
S.Em±	0.20	0.27	0.32	0.27	0.28	0.25
CD 5%	NS	0.76	0.89	0.77	0.79	0.69
Containers C						
C_1-Cloth bag	92.01 (73.58)	89.47 (71.05)	87.80 (69.59)	84.04 (66.47)	78.74 (62.56)	72.94 (58.69)
C_2 - Polythene bag 700 gauge	92.02 (73.57)	89.97 (71.53)	89.67 (71.27)	86.77 (68.69)	81.73 (64.71)	76.76 (61.19)
Mean	92.02 (73.58)	89.71 (71.32)	88.72 (70.43)	85.40 (67.58)	80.24 (63.63)	74.85 (59.94)
S.Em±	0.08	0.01	0.13	0.09	0.12	0.10
CD 5%	NS	NS	0.38	0.24	0.34	0.29

The poor seed infestation with castor oil, neem oil, sweet flag powder seed treatments may be attributed to their ovicidal oviposition inhibitory repellent and unpleasant odourus properties apart from interfering in respiratory mechanism of the storage insect pests causing greater mortality of the pulse bruchid during storage. These results are in agreement with findings of Shiva Singal (1992), Merwade (2000) and Patil (2000) in chickpea, Deshpande *et al.* (2004) in blackgram and Shivanna *et al.* (1994) in redgram.

References

Biradar, S. 2001. Influence of pre-harvest insecticidal spray on seed yield and quality and post harvest seed treatment on storability of greengram *{Vigna radiata* (L.) Wilczek.). M.Sc. *(Agri) Thesis,* University of Agricultural Sciences, Dharwad.

Deshpande, V.K., V.G. Sibi and B.S. Vyakaranahal. 2004. Effect of botanical seed treatment against *Callosobruchus chinensis* (Linn.) on viability and vigour of blackgram seeds during storage. *Seed Res.,* 32 (2): 193-196.

Merwade, M.N. 2000. Investigations on seed production techniques and storability of chickpea *(Cicer arientinum* L.). *Ph.D. Thesis,* University of Agricultural Sciences, Dharwad.

Patrick, J., V. Palanisamy and V. Vakeswaran. 1998. Influence of pre-harvest sprays on seed yield and quality of pea *(Pisum sativum* L.). *Seed Tech. News,* 28 (4):8.

Shiv Singal, K. 1992. Treatment of stored chickpea, *Cicer arietinum* L. with plant edible oils for protection against pulse beetle, *Callosobruchus chinensis* (L.). *Bull. Grain Tech.,* 30 (1): 85-87.

Shivanna, S., S. Lingappa and B.V. Patil. 1994. Effectiveness of selected plant materials as protectants against pulse beetle during storage of redgram. *Karnataka Journal of Agricultural Sciences,* 7 (3): 285-290.

— xxx —

29

Effect of Various Seed Coatings With Certain Soil Types Against *Callosobruchus chinensis* Linn. in Redgram

MUTHUKUMARAN, N[1]. AND P. THANGAVEL[2]
[1] Lecturer, Department of Entomology, [2] Reader, Department of Agricultural Botany
Faculty of Agriculture, Annamalai University, Annamalainagar - 608 002

SUMMARY

Results on the effect of different types of soils given as coating to red gram seeds against *Callosobruchus chinensis* showed that red soil effected highest mortality (85%) followed by clay and sandy soils with 70% and 65% mortality respectively. Germination test, revealed the red soil coated redgram had maximum root and shoot length. Red soil treatment protected the red gram seeds from the *C. chinensis* infestation in storage without any damage to germination.

Keywords: Seed Coating, Soil Types, Red Soil.

1. INTRODUCTION

Among the pulses, redgram (*Cajanus cajan* L.) is an important and rich proteinacious crop. Several insect pests cause reduction of quality and quantity of redgram in storage, bring in unconsumable conditions and the seeds emit bad odour, Among them *C. chinensis* plays a vital role. In general, various methods including chemicals are being followed to manage the pest population in storage. Some chemical options pose health hazards to the human beings and animals due to their residues. Farmers living in rural areas are away from the scientific advancements are practicing several indigenous methods for the management of insect pests in storage. Use of red soil as coating for the protection of redgram from pests is one such important indigenous practice. Keeping this in view, various soil types were tested against *C. chinensis* by seed coating method in red gram.

2. MATERIALS AND METHODS

Certain soil types namely red, clay and sandy were gathered and the redgram seeds were treated with them. The treatment of seeds with soils was in the ratio of 1:2. After the treatment, the redgram seeds were allowed to shade dried and kept inside the plastic container. The experiment was replicated thrice and twenty adult beetles were released to each replication. An untreated control was also maintained as check. Mortality of the adult beetles was recorded from 1, 2 and 3 days after treatment respectively. Germination percentage and root and shoot length of treated seedling were also assessed by the blotter paper method.

3. RESULTS AND DISCUSSION

Results revealed that, red soil had given maximum protection to the seeds by causing 85 per cent mortality of the pest followed by clay and sandy soils with 70 and 65 per cent mortalies respectively. It was observed that the pest mortality of the beetle in red soil treated redgram increased gradually from first day onwards. Similarly, Krishnaprasad and Vanajaram Prasad (2006) observed that the red soil coating not only acted as a physical barrier and also blocked respiratory activity of the storage pests. But in the clay soil the trend was inversely proportional to the days after treatment. The first day mortality rate was higher but decreased consequently for the next two days, whereas, in the sandy soil high mortality rate (45%) occurred on II day (Table 1).

Though in all the treated redgram seeds germination percentage was high, yet, the red soil treated redgram showed 100 percent germination followed by clay (85) and sandy (90). Among the treatments, red soil treated redgram had encouraged growth of root and shoot of seedlings (Table 2).

Table 1 : Effect of seed soil coating with various soil types agains *C. chinensis* in redgram

S. No.	Types of soil used	Percent Insect mortality at DAT			Total Insect mortality (%)
		I	II	III	
1.	Red soil	20.0 (26.56)b	30.0 (33.31)b	35.0 (36.27)a	85 (67.21)a
2.	Clay soil	35.0 (36.27)a	20.0 (26.56)c	15.0 (22.78)c	70 (56.78)b
3.	Sandy soil	0.0 (0.0)c	45.0 (42.13)a	20.0 (26.56)b	65 (53.72)c
4.	Untreated control	0.0 (0.0)d	0.0 (0.0)d	0.0 (0.0)d	0.0 (0.0)d
	CD (p = 0.05)	1.91	1.73	1.63	2.32

Each value is a mean of three replications
Values in parentheses are sine transformed
Mean values followed by a common letter are not significantly different at 5% level (DMRT)

Table 2 : Impact of various soil coatings on the seed germination and root and shoot length of redgram seedlings

S. No.	Types of soil used	Germination percentage	Growth of seedling (Length of Root and shoot) (cm)
1.	Red soil	100	14.8
2.	Clay soil	85	12.2
3.	Sandy soil	90	12.0

Each value is a mean of three replications

From the above it is concluded that, use of red as soil coating to the red gram seeds will be an measure to check the pulse beetle infestation in storage.

Reference

Krishnaprasad, G. and Vanajaramprasad, 2006. Disseminating traditional seed storage practices through people's participation. *Leisa India*, 8(3): 34-35.

— xxx —

30 Tactics in Rodent Management

HARIPRASAD, Y.[1], R. KANNAN[2] AND R.AYYASAMY[2]
[1]Professor, [2]Lecturers, Department of Entomology, Faculty of Agriculture, Annamalai University, Annamalai Nagar - 608 002

SUMMARY

This review throws light on problems and prospects of managing rodent menace in store godown.

Key words : Bait Shyness, Anticoagulant, Chronic Poison.

1. INTRODUCTION

Rodents are one of the most important notorious, non - insect pests of agricultural crops and different commodities in storage. They are ubiquitous, polyestrous and polycocous. According to Panse committee's report about 2.5% losses are caused due to the rodents in storage annually. Rodents not only feed on grain but also contaminate 20 times more than what they consume with their droppings, urine and body hair (Khare, 1990). In India, the rodents are estimated to cause 5 to 10% losses in rice. Rodents in general are prolific breeders with a gestation period of around 21 days with a size up to 20 young ones per female. A female may reproduce up to 4 times a year, averaging 6 rats per litter. The potential number of rats produced by one pair and their off spring in one year will be more than 500. Severity varies with season, location and ecosystem. The re-emergence of the rodents as an important agricultural pest in Asia (Singleton *et al.*, 2003) and the need for environmentally-sensitive approaches to management.

2. IDENTIFYING A RODENT PROBLEM

People do not often see rats, but signs of their presence are easy to detect. It is imperative to identify the species, behavior, burrow structure and activity in crops. The following signs, often evident before rodents are even seen, indicate that rodents are present.

- *Faecal droppings:* Droppings are produced in large numbers and will dry in one or two days. Rat droppings often contain hair.
- *Urine droplets:* Urine deposited as small droplets dribbles or as a line of droplets. Dried rat urine stains may not be visible to the naked eye, but will fluoresce blue-white under ultraviolet illumination.
- *Burrows or holes:* Burrows are found in the earth around field, trees, bunds, buildings, and holes in the field. Nesting and access burrows are usually less than 1.2m deep. The rats fill in the burrow entrances and check for excavation on the following day to determine if they are in use. Sounds will be heard mainly at night unless the rodent colony is overpopulated.
- *Odour:* Rats have little odour unless the population is large or well established. Mice can often be detected by their musky odour.

3. MANAGEMENT OF RATS (Mohan Rao, 1990)

3.1 Cultural Practices

Sanitation

Sanitation must remain in the forefront of any rodent control, inside and outside the building, including removal of vegetation and rubbish outside close to the building

Tidying the store, inside removing items that do not belong there and keeping strips near walls clear.

Regular cleaning by sweeping and removing all food scraps (i.e., from feeding pets or animal food in poultry sheds) in the late afternoon.

Checking incoming and outgoing produce for infestation; occasionally clearing the entire store of all produce and cleaning it out thoroughly.

Sanitation is fundamental to rat control and must be continuous. If sanitation measures are not properly maintained, the benefits of other measures will be lost, and rats will quickly return. Good housekeeping in and around buildings will reduce available shelter and food sources.

Neat, off-the-ground storage of pipes, lumber, firewood, crates, boxes, gardening equipments, and other household goods will help reduce the suitability of the area for rats and will also make their detection easy.

Garbage, trash, and garden debris should be collected frequently, and all garbage receptacles should have tight-fitting covers. For roof rats in particular, thinning dense vegetation will make the habitat less desirable.

Climbing hedges on fences or buildings are very conducive to roof rat infestations and should be thinned or remove if possible, the overhanging tree branches within 3 feet of the roof.

Separate the canopy of densely growing plants such as pyracantha and juniper from each other and from buildings by a distance of 2 feet or more to make it more difficult for the rats to move between them. Food, water and shelter are the basic needs of the rats and their elimination will restrict the rat population.

3.2. Mechanical measures

3.2.1. Trapping

The method most commonly used in buildings is trapping. Often local traps are available and in some cultures people are very good at using them. They should be placed where the rats move regularly. If placed along a wall, the trap should be perpendicular to it and the treadle with the bait should face the wall. There are several types of rodent traps including snap traps, multicatch traps, single catch live traps, and glue boards.

3.2.1.1 Snap Traps

Cheap, effective, and easy to use, snap traps are still among the most popular rat traps around. For best results, snap traps should be "pre-baited" without setting the bait for several days; and then when the rats overcome their initial fear and begin to take the bait, re-bait them and set them all on the same evening.Snap traps should be set with the trigger end in the rats' travel paths, and only in areas inaccessible to children, pets, and non-target animals.

Still, the best all-around trap for both mice and rats is the trusty snap trap. Modern snap traps have expanded plastic triggers proven to catch more rodents than older traps with smaller, metal triggers. Correct use of snap traps begins with proper placement. The snap traps should be placed along walls (rodents tend to run along walls), perpendicular to the wall, with the trigger end against the wall. They can also be placed in tandem (back-to-back), parallel to the wall so that rodents traveling in either direction will encounter the triggers. Although unbaited snap traps catch rodents, they work best when baited with food attractive to the rodents. Bait with crunchy peanut butter, nuts, meats, fruit or sweets. For mice, cotton balls and dental floss are also attractive as nesting materials. Dental floss is also advantageous for tying baits onto snap trap triggers to keep the rodents from stealing the bait. For rats, use rat-size snap traps.

While snap traps are useful in many situations, large numbers of them must be used to control larger rodent infestations. Snap traps should be placed at 10-foot intervals for mice and at 20-feet for rats. Half a dozen snap traps will capture a couple of mice in someone's kitchen, but two dozen may be required for a typical restaurant storage room, and many more in a warehouse. Thus the use of snap traps is generally too labor-intensive and time-consuming to be practical against large infestations.

3.2.1.2. Sticky or glue traps

Sticky or glue traps are another way of catching rats and mice, (Prakash, 1990, Meehan, 1984). They are boards made of wood, hard- or cardboard covered with very -sticky material. There are different types of glue available and they should be checked for suitability (stickyness, and usability in humid or dusty conditions) before large quantities are ordered. The boards are placed in the same way as traps, and normally there is no need for bait to attract rats. These traps should be checked daily, but are not regarded as very 'humane'.

The use of glue traps (glue boards) should be limited. This type of trap consists of a sticky film of glue applied to a backing of cardboard, wood or plastic. These traps can fail when they get dirty, too hot or too cold. To help keep them free of dirt and moisture, glue trap covers can be used, or the traps can be placed in boxes with openings, in empty bait stations, etc. Nevertheless, savvy rodents will avoid them, vault over them or place debris on them to cover the sticky surface.

Despite the many problems associated with their use, glue traps do catch the rodents and can play a part in an integrated rodent management program.

3.2.1.3 Live traps

Usually trap animals are unharmed so they can be released elsewhere (although sometimes the animals may injure themselves by trying to get out of the traps). Unlike the other traps on this page, live traps can *also* be used outdoors and pose little risk to non- target animals. Single catch live traps are rodent-sized cage traps of various styles. These traps capture the rat or mouse alive and unharmed, but the captured rodent must be dealth with.

3.2.1.4 Electrocution traps

They are a newer, cleaner, and more humane way to non-chemical control of the rats. The traps are baited with a suitable bait (peanut butter, cheese, oats, salami, nesting materials, etc.) and placed in the rats' travel paths. When a rat enters the trap, it immediately receives a lethal electric shock. The traps that kill rats by electrocution are available on the market. These traps are considerably more expensive than the common snap trap and can be used safely and effectively only in limited situations. Like the snap traps, these traps catch only one rat at a time and then must be emptied.

4. FUMIGATION

Fumigation should only be resorted to when all other methods have failed. The bodies of the animals remain in inaccessible locations, where their decay is likely to cause health problems. Fumigants may also damage building contents

Animals trapped in the burrows, or inhabiting air-tight buildings may be killed with fumigants such as methyl bromide, chloropicrin or calcium cyanide. These may only be used by the registered pest control organizations.

Carbon dioxide fumigation, in a sealed container, may be used to dispose of animals caught alive in traps, which are not intended to be released. It is regarded as the most humane method of killing the animals

5. CHEMICAL CONTROL

A rodenticide must have three ideal attributes namely toxicity, acceptability and safety in use.

5.1 Acute Rodenticides

Acute poisons are better for giving a quick knockdown, very lethal; ingestion leads to death of an animal within a few hours but they have little selectivity and poor efficacy.They require prebaiting and rodents exhibit bait shyness. They are single dose poisons like zinc phosphide, barium carbonate, redsquil. When zinc phosphide is ingested, it reacts with stomach acids and causes release of poisonous phosphine gas. This leads to nausea, vomiting, pulmonary edema, and eventual death.

5.2 Anticoagulant Rodenticides

Another group of rodenticide compounds are the *anticoagulants,* as far as efficacy and safety concerned they are slow in action, more laborious and management cost is more. Death from these compounds results from internal bleeding and can take from 1 - 10 days after the initial poisoning. The first anticoagulant rodenticides include *warfarin, diphacinone,* and *chlorophacinone.* Their inherent toxicity is lower than acute poisons and an antidote (vitamin K) is available in the event of accidental poisoning. Newer anticoagulants, like *brodifacum* and *bromadiolone* came into use in the 1970's. There are two classes of anticoagulant rodenticides, the *coumarins* and the *indandiones.* Both of them act by depressing the hepatic vitamin K dependent synthesis of substances essential to blood clotting. The permeability of capillaries is increased throughout the body, predisposing the animal to widespread internal hemorrhage. This usually occurs in rodents several days after bait ingestion, or after several feedings, although the newer baits can cause death after only a single feeding. The indandiones are different from the coumarins in that they can cause symptoms of injury leading to death before the haemorrhage occurs.

5.3 Formulations of Rodenticides

Tracking Powders

Tracking powders are used specifically around the perimeter of the rodent nest and are used in buildings or structures. They work by adhering to the fur as the rodent passes along the paths where the powder has been laid. The poisons are then ingested when the rodent cleans its fur.

5.4 Baits

5.4.1 Toxic Bait

While trapping is generally recommended for controlling rats indoors, when the number of rats around a building is high, toxic baits are needed to achieve adequate control, especially if there is a continuous reinfestation from surrounding areas.

It is important to note that some of the baits on the market may only be used in and around buildings and are not registered for use in the garden and landscape.

5.4.2 Anticoagulant Baits

Anticoagulant baits used to be thought of as relatively safe baits to use around the house and garden because they require multiple feedings to be effective. Newer anticoagulant baits, however, have been developed that only require a single feeding to be effective and are, therefore, more hazardous than the older type of anticoagulant bait to the pets and children. With all anticoagulants, death is generally delayed 2 to 6 days after a lethal dose of bait has been consumed. While both types of anticoagulant toxicants are effective against Norway and roof rats when formulated in acceptable baits, roof rats tend to need a few more feedings to obtain a lethal dose than Norway rats do. To make rats habituated to feed on particular food, prebaiing should be initiated. Next day the poisoned baits Chemical (zincphosphide) 2 parts,crushed food grains (as base) 96 parts, any edible oil 2 parts is placed similar manner and on the same locations instead of plain baits For control rats in burrows, poisoned baits should be placed in paper cones and pushed in to the holes without any prebaiting. This is called 'torpedo or hole baiting'. Anticoagulant baits are available in three formulations as 1. Ready to feed,2 dry baits,3water baits like in *meal,* pellet, kernel, or block form. Pelleted or meal baits are frequently sold in small packets called "place packs." Also, a wide

Descriptions and Application of Registered Rodenticides (Khare, 1990).

S. No.	Name of the chemical	Formulation	Mode of action	Recommendation
1	Acute RodenticideZinc Phosphide	Waxed granular bait	When bait is ingested, phosphine gas is generated on contact with the mucous membranes resulting in poisoning.	5 grams bait per station.
2.	Hydroxy coumarin (anticoagulant) Brodifacoum	Pellets in place packs	Single-feed anticoagulant	2 grams per rat ,0.5 grams per house mouse.
3.	Indanedione (anticoagulant) Diphacinone	Weather-resistant pellets	Multiple-feed anticoagulant causes inhibition of formation of prothrombin and reduction of clotting of blood. Mice may feed several times before effects are noticed and several days may be required for death to occur.	
4.	Chlorophacinone	Paraffin blocks, pellets, or ready-to-use grain bait Paraffin blocks and pellet formulations are moisture- resistant. Ready-to- use formulation requires no prebaiting.	Multiple-feed anticoagulant	
5.	Coumarin (anticoagulant) Bromadiolone	Maki is sold in various formulations: paraffin pellets, rat and mouse bait packs, mini blocks, rat and mouse meal bait (packs also available), and paraffin blocks. Paraffin formulations are moisture and mold resistant.	Single-feed anticoagulant	
6.	Coumarin (anticoagulant) Warfarin	Pellets, paraffin blocksControl of Norway rat, roof rat, and house mouse	Anticoagulent	
7	Repellent Tetramethylthirum disulfide (thiram)	White latex paint to be applied to base of trees, shrubs, structures,etc. (The product is applied like paint. Tree trunks and lower branches are thoroughly covered when trees are dormant in the late fall. Once, dry the "white coat" can withstand weathering for one winter season.)	Thiram repels rodents by taste only Laboratory studies' have shown Rodent Repellent to have' an acute oral LD50 of 4600 mg/kg for rats.	

variety of paraffin block baits are marketed. Most of them must be placed in a tamper-resistant bait station to prevent dogs and other animals from chewing on them.

5.4.3 Non-anticoagulant Baits

Non-anticoagulant baits are usually lethal after the rat ingested an adequate amount of toxin. Bromethalin, cholecalciferol, and zinc phosphide are common active ingredients. These baits are placed in runways or *next* to burrows where rats will find them. Nonanticoagulant baits are generally more rapid acting than anticoagulant baits and dead rodents may be found within 12 hours of baiting. Because the toxins in these baits do not accumulate in the tissue of the rodent, predators or scavengers such as dogs and cats are not likely to be adversely affected by eating the poisoned carcasses.

The ultimate success depends on the ability to provide information pertaining to technical, economical and sociological information the decision makers to early stage planning. The cooperative efforts made by the farmers, grain handlers and administrators will not provide protection unless a scheduled programming is done with a system approach. It should, thus occupy key position in operational plan for agricultural production and protection of stored grain in warehouses and storage structures.

References

Khare, B.P. 1990. Stored grain pests and their management. Kalyani publishers, Ludhiana. pp. 266 – 284.

Meehan, A.P. 1984. Rats and mice. Their biology and control. The.Rentokil library, Rentokil Limited. East Grinstead. pp. 383.

Prakash, I. 1990. Rodent control: the need for research. 'pp. 1-8 In Ed. R. Quick 'Rodents and Rice', IRRI, Manila, Philippines..

Singleton, G.R., Sudarmaji, N,P. Tuan, P.M. Sang, N.H. Huan, P.R. Brown, J. Jacob, K.L. Heong and M.M. Escalada, 2003. Reduction in chemical use following integrated ecologically-based rodent management. *International Rice Research News* 28(2): 33-35.

Singleton, G.R, Hinds, H. Leirs and Z. Zhang (Eds), 1999 Ecologically-based Management of rodent pests. Australian Centre for International Agricultural Research (ACIAR), Canberra, pp. 17-29.

— xxx —

31

Store Grain Insect Pests of Quarantine Significance for India

MURALI, R., GYANESHWER BANCHHOR
D.K. MURTHY AND N. SATHYANARAYANA
Regional Plant Quarantine Station, Meenambakkam, Chennai

SUMMARY

To reduce the risk of introduction of exotic plant pests into India, Government of India, in line with the International Standards on Phytosanitary Measures (ISPM), notified "Plant Quarantine (Regulation of Import into India) Order, 2003". Based on principles of Pest Risk Analysis, the exotic plant pests of concern to India are notified in Plant Quarantine (PQ) Order. More than 700 plant pests are notified and among these, 23 are of storage insect pests. The PQ order stipulates an effective phytosanitary measure that includes specific treatment protocols for some insect pests and prohibition of some commodities from specific country for a few pests where mitigation methods are not appropriate. The marks of identification and quarantine measures for the stored grain insect pests of concern to India are given.

Key words: Storage Insects, Quarantine, Phytosanitary Measures.

1. INTRODUCTION

In India, quarantine measures are regulated through '*The Destructive Insects and Pests Act, 1914*' (DIP Act). The DIP Act empowers the Government of India to issue notifications to regulate import of plants/ plant materials into India from time to time. After the establishment of WTO-SPS agreements, globalization of trade has taken place with removal of quantity restrictions and trade barriers. To facilitate globally harmonized, transparent, scientifically justifiable agricultural trade, International Plant Protection Convention (IPPC) has formulated many International Standards for Phytosanitary Measures (ISPMs). Due to these, the trade on agriculture has increased many folds all over the globe. The last decade has witnessed tremendous increase in the influx and out flow of plants/ plant materials in India. This has increased the risk of free flow of plant pests across the countries. To safeguard Indian agricultural, horticultural and forest plants from ingress of these pests, the Government of India, in line with International Standard for Phytosanitary Measures (ISPM), notified "*Plant Quarantine (Regulation of Import into India) Order, 2003*" (PQ order, issued under DIP Act. This PQ Order is based on principles of Pest Risk Analysis (PRA) and the pests of concern to India. More than 700 plant pests have been notified which include insects, bacteria, fungi, nematodes, viruses and weeds. Among them around 190 comprise of insect pests. Appropriate phytosanitary measures are stipulated in the PQ order to prevent the introduction of these pests of concern, which includes, from prohibition of few commodities from specific countries to specific treatment protocols

In the past several insect pests had entered into India and have caused and are still causing economically perceptible damage to native crops. Serious pests such as, coffee berry borer (*Hypothenemus hampei*), coconut eriophyid mite (*Aceria guerreronis*), spiralling white fly (*Aleurodicus dispersus*), *serpentine leaf miner* (*Liriomyza trifolii*) etc., were introduced from other countries into India and are causing huge economic losses. Still India is free from many insect pests of serious concern to other countries. In the PQ order, 23 such storage insect pests are notified. The storage insect pests of quarantine concern for India are listed in Table-1.

2. INSECT SPECIES OF QUARANTINE IMPORTANCE

The major insect pests of quarantine significance with special reference to stored grain pests for India, their marks of identification and quarantine measures are discussed below.

Table 1 : Insect species of quarantine importance

S.No.	Host	Scientific Name	Common Name
1	*Cereals*	*Prostephanus truncatus*	Larger grain borer
2	*Cereals*	*Sitophilus granarius*	Granary weevil
3	*Cereals*	*Sitophilus zeamais*	Maize weevil
4	*Cereals*	*Trogoderma variable*	Grain dermestid
5	*Pulses*	*Acanthoscelides obtectus*	Bean bruchid
6	*Pulses*	*Anomala cupripes*	Large green chafer beetle
7	*Pulses*	*Anomala pallida*	Large green chafer beetle
8	*Pulses*	*Apomyelois ceratoniae*	Carob moth
9	*Pulses*	*Bruchidium spp.*	Seed Bruchid
10	*Pulses*	*Bruchus lentis*	Bruchid
11	*Pulses*	*Melanagromyza chalcosoma*	Pod fly
12	*Pulses*	*Specularis impressithorax*	Bruchid
13	*Pulses*	*Spermophagus spp.*	Seed bruchids
14	*Oil seed*	*Stator pruininus*	Seed bruchid
15	*Oil seed*	*Trogoderma versicolor*	Dermestid beetle
16	*Coco*	*Conopomorpha cremerella*	Cocoa moth
17	*Coco*	*Hypothenemus obscurus*	Tropical nut borer
18	*Coffee*	*Hypothenemus hampei*	Coffee berry borer
19	Coco	*Acorocercops cramerella*	Coffee moth
20	*Coco*	*Hypothenemus obscurus*	Tropical nut borer
21	*Coco*	*Sahlbergia singularis*	Mirid
22	*Coco*	*Sahlbergiella theobroma*	Cocoa capsid
23	*Sheanut*	*Ephestia elutella*	Chocolate moth

2.1. Larger grain borer, *Prostephanus truncates* (Coleoptera, Bostrichidae)

Identification

The adult has the typical cylindrical Bostrichidae shape. The declivity is flattened and steep and has many small tubercles over its surface. The limits of the declivity, epically and laterally, are marked by a carina. Antennae are 10-segmented and have a loose three-segmented club; the 'stem' of the antenna is slender and clothed with long hairs and the apical club segment is as wide as, or wider than, the preceding segments. The body is 3-4.5 mm long.

Quarantine Measures

- Import of maize seeds for sowing should be free from *P. truncatus* and subject to prior approval of Department of Agriculture & Cooperation (DAC).
- Import of maize grains for consumption shall accompany PSC with an additional declaration for the freedom from *P. truncatus* and with special condition that consignment is fumigated with methyl bromide @ 32 gm/cu.m for 24 hrs. at 28^{0}C & above under NAP & the treatment shall be endorsed in Phytosanitary Certificate (PSC).

2.2. Granary weevil, *Sitophilus granarius* (L.) (Coleoptera, Curculionidae)

Identification

Adults are 2.5-5.0 mm long having the characteristic rostrum and elbowed antennae characteristic of the family Curculionidae. They are uniformly polished chestnut-brown or reddish-brown to shiny black in colour. The prothorax has distinctly oval shaped punctures. Adults do not have hind wings and therefore cannot fly.

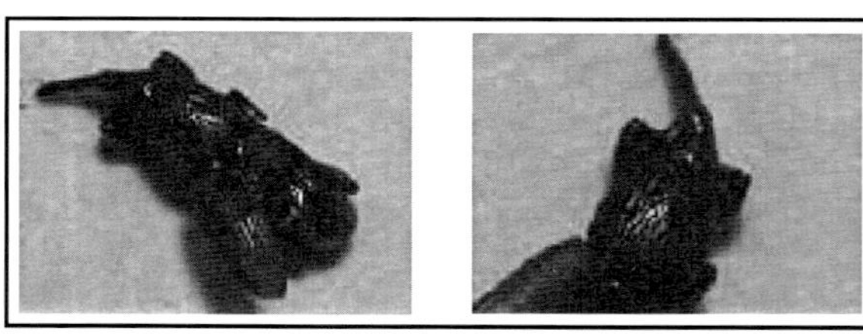

Quarantine Measures

- Import of rice, wheat & barley seed for sowing should be free from *S.granarius.*
- Consignment of rice seed for sowing shall be soaked overnight and treated in hot water at 52^{0}C for 10 min. subject to the supervision, monitoring and testing by Director, National Beurau of Plant Genetic Resources, New Delhi/ Project Director, Directorate of Rice Research, Hyderabad.
- Rice & wheat consignment for consumption shall accompany PSC with an additional declaration for the freedom from *S. granarius* and with special condition that the consignment is fumigated with methyl bromide @ 32 gm/m^3 for 24 hrs. at 28^{0}C

& above under NAP & the treatment shall be endorsed in PSC or by any other fumigant/substance in the manner approved by the Plant Protection Advisor to the Government of India for this purpose.

2.3. Maize weevil, *Sitophilus zeamais* Motschulsky (Coleoptera, Curculionidae)

Identification

Usually blacker than *S. oryzae*, females with lateral lobes of the Y-shaped sclerite pointed and their separation is greater than for *S. oryzae*. This species can be differentiated separated from *S. granarius* by the presence of wings beneath the elytra and having circular, rather than oval, shaped punctures on the prothorax.

Quarantine Measures

- Import of maize seeds for sowing should be free from *S.zeamais* & subject to prior approval of DAC.
- Import of maize grain for consumption shall accompany PSC with an additional declaration for freedom from *S. zeamais* and with special condition that the consignment is fumigated with methyl bromide @ 32 gm/cu. m for 24 h. at 28^0C & above under NAP & the treatment shall be endorsed in PSC.

2.4. Grain dermestid, *Trogoderma variable* (Coleoptera, Dermestidae).

Identification

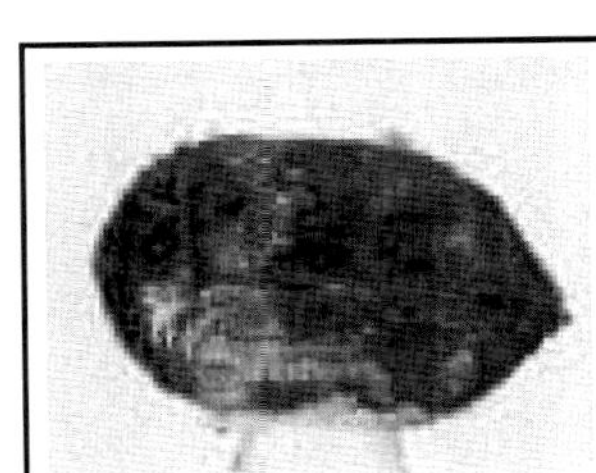

The adults resemble *T. granarium* In shape and size,.The dorsal surface is moderately clothed with fine hairs. A median ocellus is present between the compound eyes. Number of antennal segments is usually 11 and there is a fairly distinct antennal club consisting of 6-7 segments in males and 4-5 segments in females. The antennae fit into ventral grooves in the prothorax. (Crop Protection Compendium, 2003).

Quarantine Measures

Import of oat grains for consumption from Australia shall accompany PSC with an additional declaration for freedom from variable and with special condition that the consignment is fumigated with methyl bromide @ 80 gm/ m^3 for 48 h at 21^0C & above or any other treatment duly approved by the Plant Protection Adviser to the Government of India & the treatment shall be endorsed in PSC.3

2.5. Bean bruchid, *Acanthoscelides obtectus* (Coleoptera, Bruchidae)

Identification

The adult is 3-4.5 mm long having grey brown and reddish-brown body colours. Antennae are dark grey except for segments 1-5 and 11, which are reddish. The abdomen is reddish. Each hind femur has, on the inner ventral ridge near the apex, a large tooth followed by two or three smaller, sharp teeth.

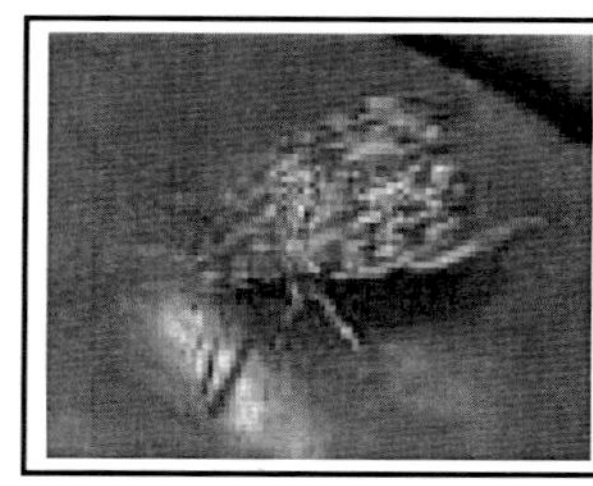

Quarantine Measures

Import of beans (*Vigna/Phaseolus*) grains for consumption shall accompany PSC with an additional declaration for freedom from *A. obtectus* and with special condition that the consignment is fumigated with methyl bromide @ 32 gm/c m^3 for 24 hrs. at 28^0C & above under NAP & the treatment shall be endorsed in PSC.

2.6. Cocoa moth, *Conopomorpha cramerella* (Lepidoptera, Racillaridae)

Identification

The adult is a small brown moth, 7 mm in length. It has a wing span of about 12 mm characterized by bright yellow patches at the tips of the forewings. The moths have very long antennae which are swept backwards in their natural resting position. Moths in flight have an appearance similar to that of large, slow-flying mosquitoes.

Quarantine Measures

Import of Rambutan (*Nephelium lappaceum*) fruits for consumption from Thailand shall accompany PSC with an additional declaration for the freedom from *cramerella* and with special condition that the consignment is fumigated with mthyl bomide @ 32 m^3 or 24 hrs. at 21^{0}C & above under NAP & the treatment shall be endorsed in PSC.

2.7. Tropical nut borer, *Hypothenemus obscurus* (Coleoptera, Scolytidae)

Identification

The adult females are 1.2-1.4 mm long, and about 2.3 times as long as wide. The colour of the female adult is light immediately after emergence, but darkens within a day to a very dark-brown. The males are approximately half the size of the females and are very rare.

Quarantine Measures

Import of cocoa bean (fermented & dried) for processing or industrial use shall accompany PSC with an additional declaration for freedom from *obscurus* and with special condition that the consignment is fumigated with methyl bromide @ 16 m^3 for24 hrs. at 21^{0}C & above under NAP & the treatment shall be endorsed in PSC.

2.8. Coffee berry borer, *Hypothenemus hampei* (Coleoptera, Scolytidae)

Identification

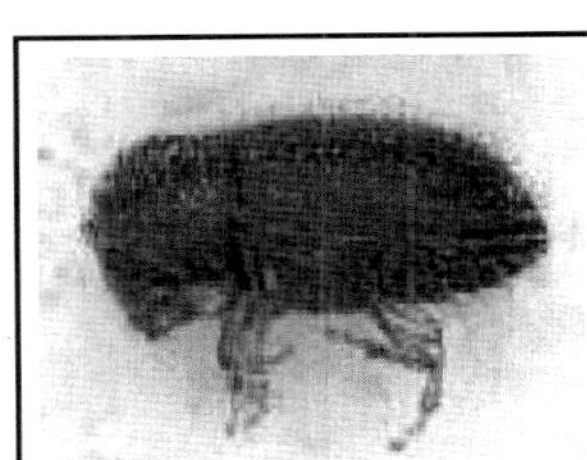

Males are often apterous, stunted and deformed. Females with body 1.4-1.6 mm long and 2.3 times as long as wide, entirely black. Antennal funicle usually 5-segmented. Pronotum with fine, raised basal and posterolateral marginal bead, anterior margin bearing 4-8 coarse teeth of about equal size. Elytra with declivity convex, gradual, extending almost to middle of elytra.

Quarantine Measures

Import of beans (seeds)/ berries(freshly harvested)/grafts/bud wood/seedlingts/rootedcuttings of coffee (Coffea spp.) is prohibited from Africa and South America.

Beans (seeds)/ berries (freshly harvested coffee (*Coffea* spp.) and related species of Rubiaceae can be imported with additional declaration for freedom from *hampei* and post entry quarantine for one year period subject to the recommendation, supervision, monitoring and testing by Director, Central Coffee Research Institute, Chikmagalur (Karnataka).

2.9. Chocolate moth, *Ephestia elutella* (Lepidoptera, Pyralidae)

Identification

The wingspan is 16-19 mm. The head and thorax are greyish-brown with the abdomen usually paler. The fore wing is greyish-brown, variably suffused with darker brown. The postmedian and antemedian fascias are dark greyish-brown; the postmedian fascia is double, with a pale line dividing it. The hind wing is white.

Quarantine Measures

Import of cocoa beans (fermented & dried) for processing or industrial use shall accompany PSC with an additional declaration for freedom from *E. elutella* and with special condition that the consignment is fumigated with methyl bromide @ 16 m^3 for 24 hrs. at 21^0C & above under NAP & the treatment shall be endorsed in PSC.

All the consignment of agricultural commodities into India, shall be inspected at the port of entry by the plant quarantine officials and they shall carry out appropriate inspection methods. In case of interception of insect pests of non-quarantine pests, the consignment shall be fumigated and released. If quarantine pests are intercepted, the consignment shall be fumigated.

Plant quarantine acts as filter rather as a barrier in the international trade. The stringent Phytosanitary Measures / Regulations are in vogue to prevent the introduction of exotic pests into India. Constant vigil and rapid quarantine screening are the only way to safeguard our agriculture from exotic insect pests.

Acknowledgement

The authors are grateful to Dr. P.S. Chandurkar, Plant Protection Adviser to the Govt. of India and Dr. Ravi Prakash, Joint Director (PP), Directorate of Plant Protection Quarantine and Storage, NH IV, Faridabad, Hariyana-121001 for constant encouragement and guidance.

References

Crop Protection Compendium 2003. Compendium Agricultural Botany International, London.

Plant Quarantine (Regulation of import into India), 2003. Ministry of Agriculture, Government of India, New Delhi.

— xxx —

Subject Index

D

E

F

G

H

I

J

K

L

M

N

S

T

V

W

Z